This series aims at speedy, informal, and high level information on new developments in mathematical research and teaching. Considered for publication are:

1. Preliminary drafts of original papers and monographs

2. Special lectures on a new field, or a classical field from a new point of view

3. Seminar reports

4. Reports from meetings

Out of print manuscripts satisfying the above characterization may also be considered, if they continue to be in demand.

The timeliness of a manuscript is more important than its form, which may be unfinished and preliminary. In certain instances, therefore, proofs may only be outlined, or results may be presented which have been or will also be published elsewhere.

The publication of the *"Lecture Notes"* Series is intended as a service, in that a commercial publisher, Springer-Verlag, makes house publications of mathematical institutes available to mathematicians on an international scale. By advertising them in scientific journals, listing them in catalogs, further by copyrighting and by sending out review copies, an adequate documentation in scientific libraries is made possible.

Manuscripts
Since manuscripts will be reproduced photomechanically, they must be written in clean typewriting. Handwritten formulae are to be filled in with indelible black or red ink. Any corrections should be typed on a separate sheet in the same size and spacing as the manuscript. All corresponding numerals in the text and on the correction sheet should be marked in pencil. Springer-Verlag will then take care of inserting the corrections in their proper places. Should a manuscript or parts thereof have to be retyped, an appropriate indemnification will be paid to the author upon publication of his volume. The authors receive 25 free copies.

Manuscripts in English, German or French should be sent to Prof. Dr. A. Dold, Mathematisches Institut der Universität Heidelberg, Tiergartenstraße or Prof. Dr. B. Eckmann, Eidgenössische Technische Hochschule, Zürich.

Die *„Lecture Notes"* sollen rasch und informell, aber auf hohem Niveau, über neue Entwicklungen der mathematischen Forschung und Lehre berichten. Zur Veröffentlichung kommen:

1. Vorläufige Fassungen von Originalarbeiten und Monographien.

2. Spezielle Vorlesungen über ein neues Gebiet oder ein klassisches Gebiet in neuer Betrachtungsweise.

3. Seminarausarbeitungen.

4. Vorträge von Tagungen.

Ferner kommen auch ältere vergriffene spezielle Vorlesungen, Seminare und Berichte in Frage, wenn nach ihnen eine anhaltende Nachfrage besteht.

Die Beiträge dürfen im Interesse einer größeren Aktualität durchaus den Charakter des Unfertigen und Vorläufigen haben. Sie brauchen Beweise unter Umständen nur zu skizzieren und dürfen auch Ergebnisse enthalten, die in ähnlicher Form schon erschienen sind oder später erscheinen sollen.

Die Herausgabe der *„Lecture Notes"* Serie durch den Springer-Verlag stellt eine Dienstleistung an die mathematischen Institute dar, indem der Springer-Verlag für ausreichende Lagerhaltung sorgt und einen großen internationalen Kreis von Interessenten erfassen kann. Durch Anzeigen in Fachzeitschriften, Aufnahme in Kataloge und durch Anmeldung zum Copyright sowie durch die Versendung von Besprechungsexemplaren wird eine lückenlose Dokumentation in den wissenschaftlichen Bibliotheken ermöglicht.

Lecture Notes in Mathematics

A collection of informal reports and seminars
Edited by A. Dold, Heidelberg and B. Eckmann, Zürich

7

Philippe Tondeur

University of Illinois, Urbana/Ill.

Introduction to Lie Groups and Transformation Groups

Second Edition

Springer-Verlag
Berlin · Heidelberg · New York 1969

Library of Congress Catalog Card Number 65–26947. Title No. 7327

PREFACE

These notes were written for introductory lectures on Lie groups and transformation groups, held at the Universities of Buenos Aires and Zurich. The notions of a differentiable manifold, a differentiable map and a vectorfield are supposed known. There is an appendix on categories and functors.

The first two chapters are influenced by a paper of R. Palais [12]. In sections 5.2 and 5.3, a lot is taken from S. Kobayashi and K. Nomizu [11]. In chapter 7, S. Helgason [6] was often used. Of course, C. Chevalley [3] was constantly consulted. The bibliography orients on the various sources. A special feature of this presentation is the systematic avoidance of the use of local coordinates on a manifold. This allows the use of the presented theory with slight modifications for Lie groups over Banach manifolds. See e.g. B. Maissen [10].

June 1964 Philippe Tondeur

PREFACE

These notes were written for introductory lectures on Lie groups and transformation groups, held at the Universities of Buenos Aires and Zurich. The notions of a differentiable manifold, a differentiable map and a vectorfield are supposed known. There is an appendix on categories and functors.

The first two chapters are influenced by a paper of R. Palais [12]. In sections 5.2 and 5.3, a lot is taken from S. Kobayashi and K. Nomizu [11]. In chapter 7, S. Helgason [6] was often used. Of course, C. Chevalley [3] was constantly consulted. The bibliography orients on the various sources. A special feature of this presentation is the systematic avoidance of the use of local coordinates on a manifold. This allows the use of the presented theory with slight modifications for Lie groups over Banach manifolds. See e.g. B. Maissen [10].

June 1964 Philippe Tondeur

CONTENTS

The * indicates a section, the lecture of which is not necessary for the understanding of the subsequent developments.

CONTENTS

The * indicates a section, the lecture of which is not necessary for the understanding of the subsequent developments.

Chapter 1. G-OBJECTS

The first two paragraphs of this chapter are essential for all that follows, whereas paragraphs 1. 3 and 1. 4 are only required for the lecture of 2. 2 and shall not be used otherwise. For the notion of category and functor, see appendix.

1. 1 Definition and examples.

If X is an object of a category $\mathfrak{K}$, we denote by Aut X the group of equivalences of X with itself. Let G be a group.

DEFINITION 1. 1. 1 An operation of G on X is a homomorphism $\tau : G \longrightarrow \text{Aut } X$. X is called a G-object with respect to τ.

An operation of G on X is a representation of G by automorphisms of X.

Example 1. 1. 2 A G-object X in the category of sets Ens is a set X equipped with a homomorphism τ of G into the group of bijections of X. Such a homomorphism is equivalently defined by a map (denoted by the same letter)

$$G \times X \xrightarrow{\ \tau\ } X$$

$$(g, x) \rightsquigarrow \tau_g(x)$$

satisfying

a) $\tau_{g_1 g_2}(x) = \tau_{g_1}(\tau_{g_2}(x))$ for $g_1, g_2 \in G$, $x \in X$

b) $\tau_e(x) = x$ for $e \in G$, $x \in X$

The last conditions in the example 1.1.2 suggest calling an operation in the sense of definition 1.1.1 more precisely a <u>left-operation</u> of G on X. A <u>right-operation</u> of G on X will then be a homomorphism $\tau : G^o \longrightarrow \mathrm{Aut}\, X$, where G^o is the opposite group of G, i.e. the underlying set of G with the multiplication $(g_1 g_2)^o = g_2 g_1$. X is then a G^o-object. We shall generally use the word operation as synonymous for left-operation and only be more precise when right-operations also occur.

Example 1.1.3 Let G be a group. If to any $g \in G$ we assign the corresponding left translation L_g of G defined by $L_g(\gamma) = g\gamma$ for $\gamma \in G$, we obtain a left-operation of G on the underlying set of G. Similarly, the assignment of the right translation R_g of G, $R_g(\gamma) = \gamma g$ for $\gamma \in G$, to any $g \in G$ defines a right-operation of G on the underlying set of G.

Example 1.1.4 Let $\rho : G \longrightarrow G'$ be a homomorphism of groups. It defines an operation τ of G on the underlying set of G' in the following way: set $\tau_g = L_{\rho(g)}$ for $g \in G$.

One obtains similarly a right operation σ by the definition

$$\sigma_g = R_{\rho(g)} \text{ for } g \in G .$$

Example 1.1.5 Let G be a group. To any $g \in G$ we assign the inner automorphism $\mathfrak{J}_g$ induced by g, $\mathfrak{J}_g(\gamma) = g\gamma g^{-1}$ for $\gamma \in G$. This defines an operation of G on itself.

Example 1.1.6 Let H be a subgroup of the group G and consider the map $G \times H \longrightarrow G$ defined by restricting the multiplication $G \times G \longrightarrow G$. It defines a right-operation of H on the set G.

Example 1.1.7 Let the group G operate on the group G' by $\tau : G \longrightarrow \operatorname{Aut} G'$. On the set $G' \times G$ the multiplication law

$$(g_1', g_1)(g_2', g_2) = (g_1' \tau_{g_1}(g_2'), g_1 g_2)$$

$$\text{for } g_i' \in G', \ g_i \in G \ (i = 1, 2)$$

defines a group structure, the semi-direct product denoted $G' \times_\tau G$. Consider the homomorphisms

$j : G' \longrightarrow G' \times_\tau G$	$j(g') = (g', e)$	for $g' \in G'$, e neutral in G
$p : G' \times_\tau G \longrightarrow G$	$p(g', g) = g$	for $g' \in G'$, $g \in G$
$s : G \longrightarrow G' \times_\tau G$	$s(g) = (e', g)$	for e' neutral in G', $g \in G$

The sequence

$$(*) \quad e \longrightarrow G' \xrightarrow{j} P \xrightarrow{p} G \longrightarrow e$$

with $P = G' \times_\tau G$ is exact and s satisfies $p \circ s = 1_G$. Conversely an exact sequence (*) and a homomorphism $s : G \longrightarrow P$ with $p \circ s = 1_G$ (a splitting of (*)) defines an operation τ of G on G' : the automorphism τ_g of G' corresponding to $g \in G$ is the inner automorphism of P defined by $s(g)$, restricted to the normalsubgroup G'. Therefore G-groups are in (1-1)-correspondence with splitting exact sequences (*).

Example 1.1.8 A typical case of the situation just mentioned is as follows: Let V be a finite-dimensional $\mathbb{R}$-vectorspace and $GL(V)$ the group of linear automorphisms of V. Then $GL(V)$ operates naturally on V. The semi-direct product $V \times GL(V)$ is the group of affine motions of V. Note that the multiplication just corresponds to the natural composition of affine motions.

We shall only have to consider categories $\mathfrak{K}$ whose objects have an underlying set and whose morphisms are applications of the underlying sets. More precisely this means that there exists a functor $V : \mathfrak{K} \longrightarrow Ens$ which can be thought of as forgetting about the additional structure on X in $\mathfrak{K}$ and taking a morphism just as an application. To avoid endless repetitions we make the following convention: From now on we shall only consider categories of that sort and shall use the same notation X for an object X and its underlying set VX.

An operation of the group G on X defines an operation on the underlying set. More generally we have the

PROPOSITION 1.1.9 Let $F : \mathfrak{K} \longrightarrow \mathfrak{K}'$ be a covariant functor from the category $\mathfrak{K}$ to the category $\mathfrak{K}'$. An operation of the group G on $X \in \mathfrak{K}$ induces a well-defined operation on $FX \in \mathfrak{K}'$.

Proof: F defines a homomorphism Aut X $\longrightarrow$ Aut FX. By composition with the given homomorphism G $\longrightarrow$ Aut X we obtain a homomorphism G $\longrightarrow$ Aut FX, which is the desired operation of G on FX.

Remark 1.1.10 If in a given category $\mathfrak{K}$ we only consider equivalences as morphisms, we obtain a new category $\mathfrak{K}_{iso}$. Evidently proposition 1.1.9 is still valid if we are only given a functor $F : \mathfrak{K}_{iso} \longrightarrow \mathfrak{K}'_{iso}$.

If $F : \mathfrak{K} \longrightarrow \mathfrak{K}'$ is a contravariant functor, a left-operation of G on X induces a right-operation of G on FX, and a right-operation on X is turned into a left-operation on FX.

Example 1.1.11 Consider the covariant functor P : Ens $\longrightarrow$ Ens, making correspond to each set X the set PX of its subsets, to each map $X \longrightarrow X'$ the induced map $PX \longrightarrow PX'$ of subsets. Let X be a G-set. Then PX is a G-set by proposition 1.1.9. The functor P^{-1} : Ens $\longrightarrow$ Ens, having the same effect as P on objects of Ens, but assigning to a map $\varphi : X \longrightarrow X'$ the map $\varphi^{-1} : PX' \longrightarrow PX$ (inverse images of subsets), transforms the G-set X into the G^o-set PX.

Example 1.1.12 Let R be a fixed object of the category $\mathfrak{R}$. The contravariant functor $h^R : \mathfrak{R} \longrightarrow$ Ens defined by $h^R(X) = [X, R]$, $h^R(\varphi)(f) = f \circ \varphi$ for $f \in [X', R]$, $\varphi : X \longrightarrow X'$, gives for any left-operation of G on X a right-operation of G on the set $[X, R]$. If $\tau : G \longrightarrow \text{Aut } X$ is the given homomorphism, we write τ^* for the induced homomorphism of G^o into the group of bijections of $[X, R]$.

Example 1.1.13 Let Λ be a ring, $\mathfrak{M}$ the category of left- Λ-modules. A G-module X is defined by an operation of G on X by Λ-linear maps; i.e. a representation of G in X in the usual sense. By proposition 1.1.9 such a representation induces an operation of G on the set of submodules of X.

Following our convention on the categories to consider, it makes sense to speak of an element of an object X.

DEFINITION 1.1.14 An element x in the G-object X is called <u>invariant or G-invariant</u> if x is fixed under every transformation τ_g : $\tau_g(x) = x$ for all $g \in G$.

A subset $M \subset X$ is called invariant if it is an invariant element of PX under the induced G-operation (example 1.1.11), i.e. if $\tau_g(M) \subset M$ for all $g \in G$.

Exercise 1.1.15 Let X and X' be G-objects of $\mathfrak{R}$ with respect to $\tau: G \longrightarrow \text{Aut } X$ and $\tau' : G \longrightarrow \text{Aut } X'$.
$\sigma_g(\varphi) = \tau'_g \circ \varphi \circ \tau_{g^{-1}}$ for $g \in G$, $\varphi: X \longrightarrow X'$ defines an operation of G on the set of morphisms from X to X'. (Example 1.1.12 is a special case of this situation, if we consider t trivial G-operation on X!) Show that there is a suitable functor inducing this operation according to proposition 1.1.9.

1.2 Equivariant morphisms.

Let G and G' be groups and $\mathfrak{R}$ a category. Suppose X to be a G-object of $\mathfrak{R}$ with respect to a homomorphism $\tau: G \longrightarrow \text{Aut } X$, X' a G'-object of $\mathfrak{R}$ with respect to a homomorphism $\tau' : G' \longrightarrow \text{Aut } X'$.

DEFINITION 1.2.1 A <u>ρ-equivariant morphism</u> $\varphi: X \longrightarrow X'$ with respect to a homomorphism $\rho: G \longrightarrow G'$

is a morphism $\varphi: X \longrightarrow X'$ of $\mathcal{R}$ such that for all $g \in G$ the following diagram commutes

$$\begin{array}{ccc} X & \xrightarrow{\varphi} & X' \\ \Big\downarrow \tau_g & & \Big\downarrow \tau'_{\rho(g)} \\ X & \xrightarrow{\varphi} & X' \end{array}$$

If $G = G'$ and $\rho = 1_G$, we just speak of an equivariant map.

Example 1. 2. 2 If X is a G-set and X' a G'-set with the operations given as in example 1. 1. 2, then a map $\varphi : X \longrightarrow X'$ is ρ-equivariant if and only if the following diagram commutes

$$\begin{array}{ccc} G \times X & \xrightarrow{\tau} & X \\ \Big\downarrow \rho \times \varphi & & \Big\downarrow \varphi \\ G' \times X' & \xrightarrow{\tau'} & X' \end{array}$$

Example 1. 2. 3 Let $\rho : G \longrightarrow G'$ be a homomorphism of groups. If G and G' are operating on itself by left-translation as in example 1. 1. 3, then a map $\varphi : G \longrightarrow G'$ is ρ-equivariant if and only if $\varphi(g_1 g_2) = \rho(g_1)\varphi(g_2)$. Therefore ρ itself is an example of a ρ-equivariant map with respect to the left-operations.

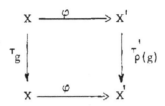

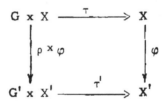

If we consider the operations of G and G' on itself by inner automorphisms, then for all $g \in G$ the diagram

$$\begin{array}{ccc} G & \xrightarrow{\rho} & G' \\ {\scriptstyle \mathfrak{J}_g}\downarrow & & \downarrow{\scriptstyle \mathfrak{J}_{\rho(g)}} \\ G & \xrightarrow{\rho} & G' \end{array}$$

commutes, i.e. ρ is ρ-equivariant.

Example 1.2.4 If we consider the right-operation of the subgroup H of G on G as in example 1.1.6, then a homomorphism $\rho : G \longrightarrow G$ sending H into H can be considered as a ρ/H-equivariant map, where ρ/H denotes the restriction of ρ to H.

Example 1.2.5 Any right-translation of a group G is an equivariant map of the G-set G defined by the left-translation. This is just the associativity law in G.

Example 1.2.6 If $\tau: G \longrightarrow \mathrm{Aut}\, X$ defines an operation of G on X, then for any $g \in G$ the map $\tau_g : X \longrightarrow X$ is $\mathfrak{J}_g$-equivariant, where $\mathfrak{J}_g : G \longrightarrow G$ denotes the inner automorphism of G defined by g.

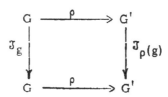

Example 1.2.7 Let X be a G-set. For fixed $x_o \in X$ $p(g) = \tau_g(x_o)$ defines a map $p : G \longrightarrow X$. If we consider the

operation of G on G by left-translation, p is an equivariant map.

If X, X', X'' are G, G', G''-objects respectively, $\rho : G \longrightarrow G'$, $\rho' : G' \longrightarrow G''$ homomorphisms and $\varphi : X \longrightarrow X'$, $\varphi' : X' \longrightarrow X''$ ρ, ρ'-equivariant morphisms respectively, then clearly $\varphi' \circ \varphi$ is a $\rho' \circ \rho$-equivariant morphism. For fixed G the G-objects of a category $\mathfrak{K}$ therefore form a category $\mathfrak{K}^G$ with the equivariant morphisms as morphisms (Definition 1.2.8).

As a complement to proposition 1.1.9 we have

PROPOSITION 1.2.9 *Let* $F : \mathfrak{K} \longrightarrow \mathfrak{K}'$ *be a covariant functor,* X, X' *respectively* G, G'*-objects of* $\mathfrak{K}$, $\rho : G \longrightarrow G'$ *a homomorphism and* $\varphi : X \longrightarrow X'$ *a* ρ*-equivariant morphism. Consider the natural operations induced on* FX *and* FX'. *Then* $F(\varphi) : FX \longrightarrow FX'$ *is* ρ*-equivariant with respect to these operations. For a fixed group* G *this defines in particular an extension of the map of proposition* 1.1.9, *sending* G*-objects of* $\mathfrak{K}$ *into* G*-objects of* $\mathfrak{K}'$, *to a functor* $F^G : \mathfrak{K}^G \longrightarrow \mathfrak{K}'^G$.

<u>Proof:</u> The commutative diagram

$$\begin{array}{ccc} X & \xrightarrow{\varphi} & X' \\ {\scriptstyle \tau_g}\big\downarrow & & \big\downarrow{\scriptstyle \tau'_{\rho(g)}} \\ X & \xrightarrow{\varphi} & X' \end{array}$$

is transformed by F in the commutative diagram

$$\begin{array}{ccc} FX & \xrightarrow{F(\varphi)} & FX' \\ {\scriptstyle F(\tau_g)}\big\downarrow & & \big\downarrow{\scriptstyle F(\tau'_{\rho(g)})} \\ FX & \xrightarrow{F(\varphi)} & FX' \end{array}$$

showing the ρ-equivariance of $F(\varphi)$ with respect to the induced operation on FX and FX'. The rest is clear.

If an equivariant morphism $\varphi : X \longrightarrow X'$ in $\mathcal{R}^G$ has an inverse $\Psi : X' \longrightarrow X$ in $\mathcal{R}$, i.e. $\Psi \circ \varphi = 1_X$, $\varphi \circ \Psi = 1_{X'}$, then Ψ is necessarily equivariant, and φ therefore an equivalence in $\mathcal{R}^G$.

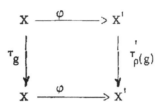

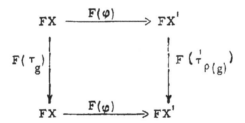

There is a canonical functor $V : \mathcal{R}^G \longrightarrow \mathcal{R}$ which consists in forgetting about the G-operation. On the other hand, we define a functor $I : \mathfrak{R} \longrightarrow \mathfrak{R}^G$ by considering on every object X of $\mathfrak{R}$ the trivial G-operation $\tau: G \longrightarrow \mathrm{Aut}\, X$ mapping G on 1_X. Therefore it makes sense to speak of equivariant morphisms $\varphi: X \longrightarrow R$, where X is a G-object of $\mathfrak{R}$ and R an arbitrary object of $\mathfrak{R}$. We call such a map an invariant morphism. More precisely we have the

DEFINITION 1.2.11 Let X be a G-object of $\mathfrak{R}$, R an object of $\mathfrak{R}$. A morphism $\varphi : X \longrightarrow R$ in $\mathfrak{R}$ is called *invariant* if for all $g \in G$ the following diagram commutes

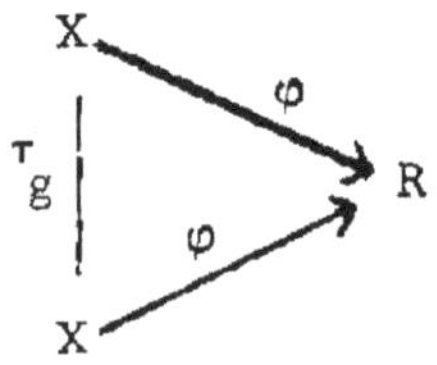

PROPOSITION 1.2.12 *Let* X *be a G-set,* X' *a* G'*-set and* $\varphi : X \longrightarrow X'$ *a* ρ*-equivariant map with respect to a homomorphism* $\rho: G \longrightarrow G'$. *If* $x \in X$ *is* G-*invariant, then* $\varphi(x)$ *is* $\rho(G)$*-invariant.*

Proof: $\tau_g(x) = x$ implies $\tau'_{\rho(g)}(\varphi(x)) = \varphi(\tau_g(x)) = \varphi(x)$.

As a consequence, G-invariant subsets of X go into $\rho(G)$-invariant subsets of X'.

Exercise 1.2.13 $\mathfrak{K}^G$ can be considered as a category of functors (interpret G as a category consisting of a single object and morphisms g with $g \in G$ with natural composition law). Equivariant morphisms are then just natural transformations. The functor F^G of proposition 1.2.9 is the canonical functor induced by F between the corresponding functor categories.

Exercise 1.2.14 If X and X' are G-objects of $\mathfrak{K}$, the set of morphisms from X to X' is a G-set according to exercise 1.1.15. The invariant elements under this operation are the equivariant morphisms $X \longrightarrow X'$. As a special case, the invariant morphisms $X \longrightarrow R$, where R is an object of $\mathfrak{K}$, are the invariant elements under the operation defined in example 1.1.12.

1.3 Orbits.

Let X be a G-set with respect to $\tau: G \longrightarrow \text{Aut } X$.

DEFINITION 1.3.1 The orbit or G-orbit of $x \in X$ under the given operation is the set $\Omega(x) = \{ \tau_g(x) / g \in G \}$.

LEMMA 1.3.2. *If X is a G-set, the different orbits form a partition of X into disjoint sets.*

Proof: As $x \in \Omega(x)$, the orbits cover X. We only have to show: if two points $x, x' \in X$ have intersecting orbits $\Omega(x), \Omega(x')$, then $\Omega(x) = \Omega(x')$. Let $y \in \Omega(x) \cap \Omega(x')$: $y = \tau_g(x), y = \tau_{g'}(x')$. For $z \in \Omega(x)$: $z = \tau_\gamma(x)$ we have $z = (\tau_\gamma \circ \tau_{g^{-1}} \circ \tau_{g'})(x') \in \Omega(x')$, i.e., $\Omega(x) \subset \Omega(x')$. This shows $\Omega(x) = \Omega(x')$.

Let X/G be the set of orbits, $\pi_X: X \to X/G$ the canonical map. An orbit is the orbit of any of its points. This implies $\pi(\tau_g(x)) = \pi(x)$, i.e. π is an invariant map. More generally we have

LEMMA 1.3.3. *Let X be a G-set, $\pi_X: X \to X/G$ the canonical map onto its orbit set X/G and R an arbitrary set. For any invariant map $\varphi: X \to R$ there is one and only one map $\Psi: X/G \to R$ such that $\varphi = \Psi \circ \pi_X$.*

Proof: If $\varphi \circ \tau_g = \varphi$ for all $g \in G$, then φ is constant on each orbit $\Omega(x)$, and therefore defines a map $\Psi: X/G \to R$ with the desired property.

On the other hand, a map $\psi : X/G \longrightarrow R$ composed with $\pi_X : X \longrightarrow X/G$ gives an invariant map $\varphi = \psi \circ \pi_X$. We have proved

PROPOSITION 1.3.4 *Let* X *be a* G-*set*, $\pi_X : X \longrightarrow X/G$ *the canonical map onto its set of orbits and* R *an arbitrary set. The correspondence* $\psi \rightsquigarrow \psi \circ \pi$, *sending maps from* X/G *to* R *into invariant maps from* X *to* R *is bijective.*

Remark. X/G is characterized by this universal property up to a canonical bijection by a standard argument. This property allows therefore the definition of X/G in an arbitrary category. Of course, there remains to show the existence of such an orbit-object in a given category.

PROPOSITION 1.3.5 *Let* X *be a* G-*set*, X' *a* G'-*set*, $\rho : G \longrightarrow G'$ *a homomorphism and* $\varphi : X \longrightarrow X'$ *a* ρ-*equivariant map. Then there exists one and only one map* $\tilde{\varphi} : X/G \longrightarrow X'/G'$, *such that the following diagram commutes*

$$\begin{array}{ccc} X & \xrightarrow{\varphi} & X' \\ {\scriptstyle \pi_X}\downarrow & & \downarrow{\scriptstyle \pi_{X'}} \\ X/G & \xrightarrow{\tilde{\varphi}} & X'/G' \end{array}$$

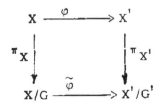

Proof: By the universal property stated in proposition 1. 3. 4, it is sufficient to show that $\pi_{X'} \circ \varphi : X \longrightarrow X'/G'$ is an invariant map. But

$$(\pi_{X'} \circ \varphi) \circ \tau_g = \pi_{X'} \circ (\varphi \circ \tau_g) = \pi_{X'} \circ (\tau'_{\rho(g)} \circ \varphi)$$

$$= (\pi_{X'} \circ \tau'_{\rho(g)}) \circ \varphi = \pi_{X'} \circ \varphi, \quad \pi_{X'} \quad \text{being}$$

an invariant map. $\widetilde{\varphi}$ is now defined as the factorization of $\pi_{X'} \circ \varphi$ through X/G.

Example 1. 3. 6   Let G be a group and H a subgroup, operating by right translations on G (example 1.1.6). Then G/H denotes the set of orbits, the set of left cosets modulo H. Let G' be another group and H' a subgroup of G'. Let further $\varphi : G \longrightarrow G'$ be a map such that $\varphi(H) \subset H'$ and $\varphi(gh) = \varphi(g)\varphi(h)$ for $g \in G$, $h \in H$. Then $\varphi/H : H \longrightarrow H'$ is a homomorphism and φ is φ/H-equivariant. By proposition 1. 3. 5 there exists one and only one map $\widetilde{\varphi} : G/H \longrightarrow G'/H'$ such that the diagram

$$
\begin{array}{ccc}
G & \xrightarrow{\varphi} & G' \\
\pi_G \downarrow & & \downarrow \pi_{G'} \\
G/H & \xrightarrow{\widetilde{\varphi}} & G'/H'
\end{array}
$$

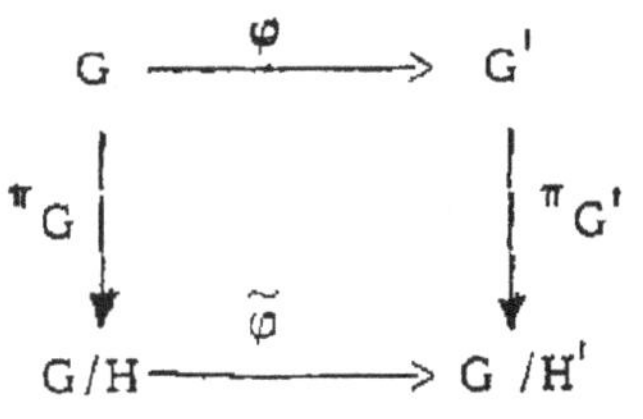

commutes. In the case where H and H' are normal subgroups of G and G' respectively and φ is a homomorphism, $\tilde{\varphi}$ is the induced homomorphism of the quotient groups.

Consider now a fixed group G. For any G-set X we have defined the orbit set X/G. Moreover by proposition 1.3.5 any equivariant map $\varphi : X \longrightarrow X'$ induces one and only one map $\tilde{\varphi}: X/G \longrightarrow X'/G$. In this way we obtain a covariant functor $B : Ens^G \rightsquigarrow Ens$ from G-sets to sets: $B(X) = X/G$, $B(\varphi) = \tilde{\varphi}$. A standard consequence is that an equivalence $\varphi: X \longrightarrow X'$ in Ens^G induces a bijection $\tilde{\varphi}: X/G \longrightarrow X'/G$.

Remark. If we consider the "forget-functor" $V : Ens^G \longrightarrow Ens$ defined as forgetting about the G-set structure, we see that $\pi : V \longrightarrow B$ is a natural transformation of V into B.

To the beginning of this paragraph, for a G-set X, we have introduced the map $\Omega : X \longrightarrow PX$, which can also be described by the map $\pi_X : X \longrightarrow X/G$ as $\Omega = \pi_X^{-1} \circ \pi_X$. The right side can be extended to a map $PX \longrightarrow PX$ and we interpret now Ω as the map $\pi_X^{-1} \circ \pi_X : PX \longrightarrow PX$. For $M \subset X$ $\Omega(M)$ is just the orbit of M under the induced G-operation on PX. Explicitly

$$\Omega(M) = \{ \tau_g(x) / g \in G, x \in M \}.$$

$\Omega(M)$ is therefore the saturation of M with respect to G, i.e. the union of all G-orbits of X intersecting M.

The invariance of $M \subset X$ can now be expressed by $\Omega(M) = M$. For an arbitrary $M \subset X$ the set $\Omega(M)$ is the intersection of all invariant sets containing M. The orbits are the minimal invariant sets.

Isotropy groups. Let X be a G-set and $x \in X$. Consider $G_x = \{g \in G / \tau_g(x) = x\}$. G_x is a subgroup of G.

DEFINITION 1.3.7. G_x is called the isotropy group of x.

PROPOSITION 1.3.8. $G_{gx} = gG_xg^{-1}$.

Proof: For simplicity we write $\tau_g(x) = gx$. Then for $h \in G_x$ we have $ghg^{-1} . gx = ghx = gx$, which implies $gG_xg^{-1} \subset G_{gx}$. As $g^{-1} . gx = x$, we have by the same argument $g^{-1}G_{gx}g \subset G_x$ or $G_{gx} \subset gG_xg^{-1}$, which proves the proposition.

This can also be expressed in the following way. Consider the map $\varphi : X \longrightarrow SG$ into the set of subgroups of G, defined by $\varphi(x) = G_x$. G operates by inner automorphisms on SG, an orbit being a conjugacy class of subgroups. By proposition 1.3.8 the diagram

$$\begin{array}{ccc} X & \xrightarrow{\varphi} & SG \\ \tau_g \Big\downarrow & & \Big\downarrow \mathfrak{T}_g \\ X & \xrightarrow{\varphi} & SG \end{array}$$

commutes for all $g \in G$, i.e. φ is an equivariant map. Therefore φ induces following proposition 1.3.5 a map $\tilde{\varphi} : X/G \longrightarrow SG/G$, i.e. to every orbit of X there corresponds a well-defined conjugacy class of subgroups of G, called the orbit-type of the orbit.

Particular orbit-types are the conjugacy classes $\{e\}$ and G. Let x_o be in $\Omega(x_o)$ of orbit-type $\{e\}$. Then for $x \in \Omega(x)$ there is one and only one $g \in G$ such that $gx_o = x$. Because $g_1 x_o = g_2 x_o$ or $g_2^{-1} g_1 x_o = x_o$ implies $g_2^{-1} g_1 = e$ and $g_1 = g_2$. If x_o is in $\Omega(x_o)$ of orbit-type G, then x_o is G-invariant, and $\Omega(x_o) = x_o$. Therefore the fixed points are exactly the orbits of orbit-type G.

Example 1.3.9. Consider the full linear group $GL(n, \mathbb{R})$, consisting of the real quadratic matrices with n entries having a determinant different from zero, with the natural operation on $\mathbb{R}^n$. The origin O and its complement $\mathbb{R}^n - \{o\}$ are the orbits of this operation. The orbit-type of O is $GL(n, \mathbb{R})$.

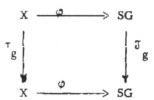

Example 1.3.10. Consider $\mathbb{R}^n$ with the standard euclidean metric and the corresponding orthogonal group $O(n, \mathbb{R})$. The orbits of the natural operation are the spheres with the origin as center. The isotropy group of a point different from the origin is isomorphic to the orthogonal group $O(n-1, \mathbb{R})$.

Example 1.3.11. Let X denote the complex plane with a point at infinity adjoined. The group of transformations of the type $z \rightsquigarrow \frac{az + b}{cz + d}$ with $a, b, c, d \in \mathbb{C}$ and $ad - bc \neq 0$ operates on X. X is the orbit of any point $x \in X$.

Example 1.3.12. Consider the operation of a group G on itself by inner automorphisms. The fixpoints are the elements of the center CG. We have already considered the induced G-operation on the set SG of subgroups of G. The orbit of a subgroup is its conjugacy class. Therefore the invariant subgroups of G are exactly the fixpoints under this operation. Moreover it follows that the different conjugacy classes form a partition of SG.

The effect of an equivariant map on the isotropy groups is described by the

PROPOSITION 1.3.13. Let X be a G-set, X' a G'-set, $\rho : G \longrightarrow G'$ a homomorphism and $\varphi : X \longrightarrow X'$ a ρ-equivariant map. Then $\rho(G_x) \subset G'_{\varphi(x)}$.

Proof: Let $g \in G_x$, i.e. $gx = x$. Then $\rho(g)\varphi(x) = \varphi(gx) = \varphi(x)$, i.e. $\rho(g) \in G'_{\varphi(x)}$.

Exercise 1.3.14. Let X be a G-set, X' a G'-set, $\rho : G \longrightarrow G'$ a homomorphism and $\widetilde{\varphi} : X/G \longrightarrow X'/G'$ a map. Study the conditions under which $\widetilde{\varphi}$ is induced by a ρ-equivariant map $\varphi : X \longrightarrow X'$ in the sense of proposition 1.3.5.

Exercise 1.3.15. For a G-set X consider the map $\Omega = \pi_X^{-1} \circ \pi_X : PX \longrightarrow PX$ defined above. Show that Ω has the following properties:

a) $\Omega(\emptyset) = \emptyset$ for the empty set $\emptyset$ of X

b) $M \subset \Omega(M)$ for $M \subset X$

c) $\Omega(\Omega(M)) = \Omega(M)$

d) $\Omega(\cup_{\lambda\in\Lambda} M_\lambda) = \cup_{\lambda\in\Lambda} \Omega(M_\lambda)$

for a family $(M_\lambda)_{\lambda\in\Lambda}$ of $M_\lambda \subset X$.

Therefore Ω is a "Kuratowski-operator" on X and defines

a topology on X according to the definition: $M \subset X$ is closed

if and only if $\Omega(M) = M$. This remains true if we consider

an arbitrary equivalence relation R on X (not necessarily

defined by a group G) and the map $\Omega = \pi_x^{-1} \circ \pi_x : PX \longrightarrow PX$,

where $\pi_x : X \longrightarrow X/R$ is the canonical map onto the quotient

set X/R. Show that more generally for an arbitrary relation

R on a set X the "saturation-operator" $\Omega : PX \longrightarrow PX$

defined by

$$\Omega(M) = \{ y \in X \big/ x R y \text{ for some } x \in M \}$$

for $M \subset X$ satisfies the properties 1) to 4) if and only if R

is a reflexive and transitive relation on X.

Exercise 1.3.16. Consider the topology defined in exercise

1.3.15 on a set X equipped with an equivalence relation R.

Show the following properties:

1) $M \subset X$ is closed if and only if M is a union of
    equivalence classes;

2) $M \subset X$ is closed if and only if M is open.

What are the conditions on X/R for the topology in question to satisfy

1) the second countability axiom,

2) to be compact,

3) to be connected?

Exercise 1.3.17. Let X be a G-set, R an arbitrary set and $\varphi : X \longrightarrow R$ a map. Suppose X equipped with the topology defined in exercise 1.3.15 and R topologized by the discrete topology. Then φ is invariant if and only if φ is continuous.

1.4 Particular G-sets.

Let X be a G-set, defined by a homomorphism $\tau : G \longrightarrow \mathrm{Bij}\, X$. We define some particular properties an operation can have.

DEFINITION 1.4.1 τ is an effective operation if τ is injective, i.e., $\mathrm{Ker}\ \tau = \{e\}$.

We observe that $\mathrm{Ker}\ \tau = \bigcap_{x \in X} G_x$, an element of $\mathrm{Ker}\ \tau$ being exactly an element of G contained in every isotropygroup. If τ is not effective, then there exists a factorization $\tilde{\tau}$ through $G/\mathrm{Ker}\ \tau$

$$\begin{array}{ccc} G & \xrightarrow{\ \tau\ } & \mathrm{Bij}\, X \\ \downarrow & \nearrow_{\tilde{\tau}} & \\ G/\mathrm{Ker}\ \tau & & \end{array}$$

and $G/\mathrm{Ker}\ \tau$ operates effectively on X.

Example 1.4.2 The operation $\mathfrak{J}$ of a group G by inner automorphisms has the center $CG = \mathrm{Ker}\ \mathfrak{J}$ as kernel.

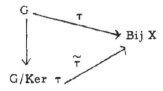

DEFINITION 1.4.3 τ is a <u>free operation</u>, if $\tau_g(x) = x$ for some $x \in X$ implies $g = e$.

This means that a transformation τ_g for $g \neq e$ has no fixpoint. Free means "free of fixpoints". The isotropy group is reduced to the neutral element: $G_x = \{e\}$ for every $x \in X$. X is also called a "G-principal set". Note that a free operation is effective.

Example 1.4.4 The operation of G on G by left-translations is free. Let H be a subgroup of G, operating on G by right-translations. This operation is free.

DEFINITION 1.4.5 τ is a <u>transitive operation</u>, if for $x_1, x_2 \in X$ there exists a $g \in G$ such that $\tau_g(x_1) = x_2$, <u>simply transitive</u>, if, moreover, the element g is unique.

A simply transitive operation is free. Conversely, a free operation is simply transitive on each orbit. Because if $x = g_i x_o$ $(i = 1, 2)$, then $x_o = g_2^{-1} x = g_2^{-1} g_1 x_o$ and therefore $g_2^{-1} g_1 \in G_{x_o} = \{e\}$, i.e. $g_1 = g_2$.

The definition of a transitive operation can also be put in the following form: there exists an element $x_o \in X$ such that $\Omega(x_o) = X$. X is then the orbit of each point $x \in X$. This shows that the set of orbits X/G is a point. This property allows us to define the transitivity of a G-operation in an arbitrary category, as soon as the notion of point is defined.

DEFINITION 1.4.6 A G-set X is called <u>homogeneous</u>, if G operates transitively on X.

Example 1.4.7 The orthogonal group $O(n, \mathbb{R})$ operates transitively on the unit sphere S^{n-1} in $\mathbb{R}^n$.

More generally, an operation of G on X defines a transitive operation on each G-orbit.

Example 1.4.8 The group of holomorphisms of the unit disk in the complex plane operates transitively.

A fundamental example of a homogeneous G-set is obtained in the following way. Consider a group G and a subgroup H operating on G by right-translations. Then we can define an operation of G on the orbit set G/H. The left translation $L_g : G \longrightarrow G$ satisfies $L_g(\gamma H) = g\gamma H$ and therefore defines by $\sigma_g(\gamma H) = g\gamma H$ a map $\sigma_g : G/H \longrightarrow G/H$. σ is the desired operation, making G/H a G-set, which is evidently homogeneous.

Remark. The isotropy group of H is H.

We shall show that for an arbitrary homogeneous G-set X there exists a subgroup H of G and an equivalence $\varphi: G/H \longrightarrow X$ of G-sets, where G/H is considered as a G-set in the sense indicated.

First let X be an arbitrary G-set and $x_0 \in X$. We put $H = G_{x_0}$ and define $\varphi : G/H \longrightarrow \Omega(x_0) \quad X$ by $\varphi(gH) = gx_0$.

LEMMA 1.4.9 φ <u>is equivariant and injective.</u>

<u>Proof:</u> For $\gamma \in G$ one has $(\tau_\gamma \circ \varphi)(gH) = \tau_\gamma(gx_0) = \gamma g x_0$ and $(\varphi \circ \tau_\gamma)(gH) = \varphi(\gamma g H) = \gamma g x_0$, and therefore $\tau_\gamma \circ \varphi = \varphi \circ \sigma_\gamma$, i.e. the equivariance of φ. To show the injectivity, consider $g_1, g_2 \in G$

such that $\varphi(g_1H) = \varphi(g_2H)$. This means $g_1x_o = g_2x_o$ or $g_2^{-1}g_1x_o = x_o$ and therefore $g_2^{-1}g_1 \in H$. But $g_1 \in g_2H$ implies $g_1H = g_2H$, q. e. d.

If X is homogeneous, then $\Omega(x_o) = X$ and φ is an equivalence. We have proved

PROPOSITION 1. 4. 10 _Let_ X _be a homogeneous_ G _-set and select_ $x_o \in X$. _Let_ H _be the isotropy group of_ x_o _and consider the_ G-_operation_ _on_ G/H _induced by the left-translation of_ G. _Then the map_ $\varphi : G/H \longrightarrow X$ _defined by_ $\varphi(gH) = gx_o$ _is an equivalence of_ G-_sets._

The group H depends on the choice of $x_o \in X$, but the conjugacy class of H is well-defined by the operation in view of the transitivity.

We conclude this chapter by some remarks on effective and transitive operations on sets. In view of the preceding proposition, we can consider without loss of generality G-sets of the type G/H, where H is a subgroup of G. The kernel K of the homomorphism defining the operation of G on G/H is the intersection of the isotropy groups, therefore

$$K = \bigcap_{g \in G} gHg^{-1} .$$

K is an invariant subgroup of G contained in H. Conversely, if L is an invariant subgroup of G with $L \subset H$, then $L \subset K$, because $lgH = gl'H$ for some $l' \in L$ in view of $Lg = gL$ and $lgH = gH$ which signifies $L \subset K$. Therefore we have

PROPOSITION 1.4.11 Let G be a group, H a subgroup and consider the G-operation on G/H induced by the left-translations of G. The kernel K of the homomorphism $\sigma : G \longrightarrow Bij(G/H)$ defining this operation is the greatest invariant subgroup of G contained in H and can be described as

$$K = \bigcap_{g \in G} gHg^{-1} .$$

COROLLARY 1.4.12 G operates effectively on G/H if and only if H contains no invariant subgroup of G different from $\{e\}$.

Exercise 1.4.13 Study the effect of the choice of the point $x_0 \in X$ in proposition 1.4.10.

Chapter 2. G-SPACES

2.1 Definition and examples.

DEFINITION 2.1.1 A topological group G is a group which is a topological space such that the maps

$$G \times G \longrightarrow G \quad , \quad G \longrightarrow G$$

$$(g_1, g_2) \rightsquigarrow g_1 g_2 \qquad g \rightsquigarrow g^{-1}$$

are continuous.

DEFINITION 2.1.2 Let G be a topological group. A G-space X is a topological space which is a G-set with respect to a map $G \times X \longrightarrow X$. Moreover this map is supposed to be continuous. The pair (G, X) is also called a topological transformation group.

It is clear that the group G is acting by homeomorphisms on X, so that X is a G-object in the category of topological spaces. We require, moreover, the continuity of the map $G \times X \longrightarrow X$.

Let G and G' be topological groups.

DEFINITION 2.1.3 A homomorphism $\rho : G \longrightarrow G'$ of topological groups is a homomorphism of groups, which is continuous.

Let X be a G-space, X' a G'-space and $\rho : G \longrightarrow G'$ a homomorphism.

DEFINITION 2.1.4 A ρ-<u>equivariant map</u> $\varphi: X \longrightarrow X'$ is a ρ-equivariant map in the sense of definition 1.2.1 which is continuous.

The map φ makes the following diagram commutative

$$\begin{array}{ccc} G \times X & \longrightarrow & X \\ {\scriptstyle \rho \times \varphi}\downarrow & & \downarrow{\scriptstyle \varphi} \\ G' \times X' & \longrightarrow & X' \end{array}$$

φ is continuous and therefore also $\rho \times \varphi$, as follows immediately by the universal property of the product topology.

An equivalence of G-spaces X, X' is an equivalence $\varphi: X \longrightarrow X'$ of G-sets which is a homeomorphism.

Example 2.1.5 Let G be a topological group. The operation of G on G by left or right-translations makes the space G a G-space. The operation of G on G by inner automorphisms also makes G a G-space.

Remark. Let X be a topological space and G the group of homeomorphisms of X. The discrete topology on G certainly makes X a G-space.

Let X be a compact G-space. Consider the group Aut X of homeomorphisms with the compact-open topology. It can be proved that Aut X is a topological group, and that the map $G \times X \longrightarrow X$ is continuous.

2.2 Orbitspace.

Let G be a topological group and X a G-space. Consider the set of orbits X/G and the canonical map $\pi_X : X \longrightarrow X/G$. The quotient topology on X/G is the strongest topology on X/G making π_X continuous. The open sets of X/G are the sets having an open saturation in X.

DEFINITION 2.2.1 The orbit space X/G of the G-space X is the set of orbits with the quotient topology.

PROPOSITION 2.2.2 $\pi_X : X \longrightarrow X/G$ is an open map. The topology on X/G is characterized as being the unique topology making the map π_X continuous and open.

Proof: Let $M \subset X$ be open. $\tau_g(M)$ is open and therefore also $\Omega(M) = (\pi_X^{-1} \circ \pi_X)(M)$, being the union of all sets $\tau_g(M)$. But this means that $\pi_X(M)$ is open by definition of the quotient topology. To prove the second statement, consider more generally a map $\varphi : X \longrightarrow Y$ from X to a set Y. Two topologies on Y making both φ continuous and open necessarily coincide. Because if O is an open set of Y in the one topology, $\varphi^{-1}(O)$ is open in X and $\varphi(\varphi^{-1}(O)) = O$ is also open in the other topology.

Example 2.2.3 Let G be a topological group and H a subgroup of G with the relative topology. The operation of H on G by right translations makes G an H-space. The canonical map $\pi_G : G \longrightarrow G/H$ onto the orbitspace is continuous and open.

The quotient topology on X/G can also be characterized by the following property. Let R be an arbitrary topological space. The map $\Psi \rightsquigarrow \Psi \circ \pi$, sending continuous maps $\Psi : X/G \longrightarrow R$ into continuous maps $X \longrightarrow R$ is injective. The proposition 1. 3. 4 can therefore now be completed by

PROPOSITION 2. 2. 4 *Let* G *be a topological group,* X *a* G*-space,* $\pi_X : X \longrightarrow X/G$ *the canonical map onto the orbitspace and* R *an arbitrary space. The correspondence* $\Psi \rightsquigarrow \Psi \circ \pi$, *sending continuous maps from* X/G *to* R *onto invariant continuous maps from* X *to* R *is bijective.*

PROPOSITION 2. 2. 5 Let G, G' be topological groups, $\rho : G \longrightarrow G'$ a homomorphism and X, X' respectively G, G'-spaces. A ρ-equivariant map $\varphi : X \longrightarrow X'$ induces one and only one continuous map $\tilde{\varphi} : X/G \longrightarrow X'/G'$ such that the following diagram commutes

$$\begin{array}{ccc} X & \xrightarrow{\ \varphi\ } & X' \\ {\scriptstyle \pi_X}\downarrow & & \downarrow{\scriptstyle \pi_{X'}} \\ X/G & \xrightarrow{\ \tilde{\varphi}\ } & X'/G' \end{array}$$

Proof: There is only to show the continuity of $\tilde{\varphi}$. But this is a consequence of the continuity of $\pi_{X'} \circ \varphi$ in view of proposition 2. 2. 4.

Exercise 2. 2. 6 Consider a G-space X with a transitive operation of G on X. Select $x_o \in X$ and let H be the isotropy group of x_o.

Define, as in proposition 1.4.10, a map $\varphi : G/H \longrightarrow X$. This map is an equivalence of G-sets and continuous, but not necessarily a homeomorphism. The following counter-example is taken from Bourbaki. Let $\mathbb{R}$ operate on $\mathbb{T}^2 = \mathbb{R}^2/\mathbb{Z}^2$ by $\tau_\lambda(x_1, x_2) = (x_1 + \alpha(\lambda), x_2 + \alpha(\theta\lambda))$ where $\lambda \in \mathbb{R}$, $(x_1, x_2) \in \mathbb{T}^2$, $\alpha: \mathbb{R} \longrightarrow \mathbb{R}/\mathbb{Z}$ the canonical homomorphism and θ an irrational number. Fixing $(x_1, x_2) \in \mathbb{T}^2$ we define $\varphi: \mathbb{R} \longrightarrow \mathbb{T}^2$ as in lemma 1.4.9, obtaining a continuous injection. Consider the image $X = \Omega(x_1, x_2)$ with the relative topology. $\varphi: \mathbb{R} \longrightarrow X$ is a continuous bijection, but not a homeomorphism. Because X is dense in $\mathbb{T}^2$ and cannot be homeomorphic to the complete space $\mathbb{R}$.

Exercise 2.2.7 Let G be a topological group and H an open subgroup. Then H is closed in G. (Consider the partition of G defined by the elements of G/H.)

Exercise 2.2.8 Let G be a connected topological group and U a neighborhood of e. The neighborhood $V = U \cap U^{-1}$ has the properties: $V \subset U$, $V^{-1} = V$. Consider the sets $V^n = \{g_1 \cdot \ldots \cdot g_n / g_i \in V, i = 1, \cdots, n\}$. The union $V^\infty = \cup V^n$ is a group, the group generated by V. e is an inner point of V^∞, as $e \in V \subset V^\infty$. Any point of V^∞ is therefore an inner point, the left-translations being homeomorphisms leaving V^∞ invariant. V^∞ is an open subgroup of G and therefore closed. As G is connected, this shows $V^\infty = G$. This proves that G is generated by an arbitrary neighborhood U of e.

Exercise 2.2.9 Let G be a topological group and G_0 the connected component of the neutral element $e \in G$, the identity component of G. Show that G_0 is a closed invariant subgroup of G.

Chapter 3. G-MANIFOLDS

This chapter introduces the fundamental notions of these lectures. In the following chapters, we proceed to a detailed study of G-manifolds and Lie groups.

3.1 Definition and examples of Lie groups.

Manifold will mean a Hausdorff, but not necessarily connected manifold.

DEFINITION 3.1.1 A *Lie group* is a group G which is an analytic manifold such that the maps

$$G \times G \longrightarrow G \qquad\qquad G \longrightarrow G$$

$$(g_1, g_2) \rightsquigarrow g_1 g_2 \qquad\qquad g \rightsquigarrow g^{-1}$$

are analytic.

Differentiable shall always mean C^∞. If one replaces analycity by differentiability in the definition above, it doesn't change anything; i.e., analycity is then automatically satisfied (Pontrjagin, [14], p. 191). For a great part of the theory, we shall only make explicit use of differentiability.

In the definition above, analytic manifold means real analytic manifold. Replacing it by complex analytic manifold, one obtains the notion of a complex Lie group.

Two arbitrary connectedness components G_1, G_2 of a Lie group G are analytically diffeomorphic. For $g_1 \in G_1$, $g_2 \in G_2$ the map $g \rightsquigarrow g_2 g_1^{-1} g$ is an example of such a diffeomorphism. All the connectedness components therefore have the same dimension and it makes sense to speak of the dimension of a Lie group.

Example 3.1.2 The additive group $\mathbb{R}^n$ or $\mathbb{C}^n$; $\mathbb{T}^n = \mathbb{R}^n / Z^n$; $GL(n, \mathbb{R})$ - the group of quadratic matrices with n rows and determinant different from zero.

Example 3.1.3 Let G be a Lie group and TG the tangent bundle. Then TG is a Lie group. This follows from the fact that T is a functor conserving direct products.

Example 3.1.4 Let G_1 and G_2 be Lie groups. Then the direct product $G_1 \times G_2$ is a Lie group.

DEFINITION 3.1.5 Let G and G' be Lie groups. A <u>homomorphism</u> $\rho : G \longrightarrow G'$ of Lie groups is a homomorphism of groups which is analytic.

Remark. It is to be noted that in the literature the term homomorphism is often reserved for analytic homomorphisms of groups such that the map $\rho : G \longrightarrow \rho(G)$ is open.

Example 3.1.6 Let V be an n-dimensional vector space over $\mathbb{R}$. The choice of a base $e_1, \ldots, e_n$ of V defines an isomorphism $GL(V) \longrightarrow GL(n, \mathbb{R})$ of groups, permitting us to define a Lie group structure on the group of linear automorphisms GL(V) of V. This structure is independent of the choice of the base. Because two choices of the base of V correspondtotwo isomorphisms $GL(V) \longrightarrow GL(n, \mathbb{R})$ which differ by an inner automorphism of $GL(n, \mathbb{R})$.

Example 3.1.7 Let G be a Lie group and TG the tangent bundle with its Lie group structure (example 3.1.3). Consider the tangent space G_e of G at the identity e of G and its natural injection $j : G_e \longrightarrow TG$. If G_e is equipped with the Lie group structure defined by addition, j is a homomorphism of Lie groups. The natural projection $p : TG \longrightarrow G$, assigning to each tangent vector its origin, is also a homomorphism of Lie groups. The sequence

$$O \longrightarrow G_e \xrightarrow{j} TG \xrightarrow{p} G \longrightarrow e$$

is exact. Moreover, there exists a splitting, the natural injection $s : G \longrightarrow TG$, satisfying $p \circ s = 1_G$.

Exercise 3.1.8 Let G be a locally Euclidean topological group, i.e., having a neighborhood of the identity e homeomorphic to an open subset of an Euclidean space. The identity component G_o of G has a countable base. Therefore G is paracompact.

Exercise 3.1.9 A Lie group is locally connected.

Exercise 3.1.10 The identity component G_o of a Lie group is an open subgroup of G.

3.2 Definition and examples of G-manifolds.

DEFINITION 3.2.1 Let G be a Lie group. A G-manifold X is a differentiable manifold X which is a G-set with respect to a map $G \times X \longrightarrow X$. Moreover this map is supposed to be differentiable. The pair (G, X) is also called a Lie transformation group.

The group G is acting by diffeomorphisms on X, so that X is a G-object in the category of differentiable manifolds. Moreover the differentiability of the map $G \times X \longrightarrow X$ is required.

Let X be a G-manifold, X' a G'-manifold, and $\rho: G \longrightarrow G'$ a homomorphism of Lie groups.

DEFINITION 3.2.2 A ρ-equivariant map $\varphi: X \longrightarrow X'$ is a ρ-equivariant map in the sense of definition 1.2.1 which is differentiable.

Example 3.2.3 $\mathbb{R}$-manifolds are of fundamental importance for the theory of G-manifolds. They have received a special name: one-parameter groups of transformations. We shall take up the study of one-parameter groups of transformations in chapter 5.

Example 3.2.4 The operation of a Lie group G on the underlying manifold by left-translations defines G as a G-manifold. The operation of G on itself by inner automorphisms also defines G as a G-manifold.

Example 3.2.5 Let V be a finite-dimensional $\mathbb{R}$-vectorspace. GL(V) is then a Lie group. Let G be a Lie group and $\tau: G \longrightarrow GL(V)$ a homomorphism. We call τ a representation of the Lie group G in V.

Remark. As observed at the end of section 2.1, for a locally compact G-space X the continuity of the map $G \times X \longrightarrow X$ can be expressed by the continuity of the homomorphism $G \longrightarrow \text{Aut } X$ defining the operation, if Aut X is equipped with the compact-open topology. One would like to describe similarly the differentiability of the map $G \times X \longrightarrow X$ for a G-space X. But for this the group Aut X of diffeomorphisms of X should first be turned into a manifold (modeled over a sufficiently general topological vectorspace), which presents serious difficulties. Nevertheless we shall use this viewpoint for heuristical remarks.

Example 3.2.6 Let X be a G-manifold and T the functor assigning to each differentiable manifold its tangent bundle. Then TX is a TG-manifold, because T conserves direct products. G being a subgroup of TG (example 3.1.7), TX is also a G-manifold. This justifies many classical notations in the theory of transformation groups, which at first sight seem abusively short.

Example 3.2.7 Let G and G' be Lie groups and G' a G-manifold with respect to an operation $\tau : G \longrightarrow \text{Aut } G'$. Then the semi-direct product $G' \times_{\tau} G$ defined in example 1.1.7 is a Lie group with the analytic structure of the product-manifold. This generalizes example 3.1.4, which corresponds to the trivial operation of G on G'.

Let V be a finite dimensional $\mathbb{R}$-vectorspace. The group of affine motions of V, which is the semi-direct product $V \times_{\cdot} GL(V)$ with respect to the natural operation of GL(V) on V, is a Lie group by the preceding.

Example 3.2.8 Let G be a Lie group and consider the exact sequence

$$0 \longrightarrow G_e \xrightarrow{j} TG \xrightarrow{p} G \longrightarrow e$$

of example 3.1.7. The splitting $s : G \longrightarrow TG$ defined by the natural injection of G gives rise to an operation of G on the additive group G_e defined by $\tau_g = \mathfrak{J}_{s(g)}/G_e$ (example 1.1.7). This representation of G in G_e plays an important role in the theory of Lie groups (adjoint representation). TG is isomorphic to the semi-direct product $G_e \times_\tau G$ with respect to this operation τ.

Chapter 4. VECTORFIELDS

In this chapter we begin with the detailed theory of G-manifolds and Lie groups. The Lie algebra of a Lie group is defined and the formal properties of this correspondence are studied.

4.1. Real functions.

The adjective "differentiable" shall be omitted from now on, it being understood that all manifolds and maps are differentiable.

Let X be a manifold and denote by CX the set of real-valued functions on X. CX is a commutative ring with identity, the operations on functions being defined pointwise. It can also be considered as an algebra over the reals $\mathbb{R}$, identifying the set of constant functions on X with $\mathbb{R}$.

Let X' be another manifold. A map $\varphi : X \longrightarrow X'$ induces a map $\varphi^* : CX' \longrightarrow CX$ defined by $\varphi^*(f') = f' \circ \varphi$ for $f' \in CX'$. φ^* is a ring homomorphism respecting identities. If we consider the $\mathbb{R}$-algebra structure on CX and CX', then φ^* is a homomorphism of $\mathbb{R}$-algebras respecting identities. This shows that the correspondence $X \rightsquigarrow CX$, $\varphi \rightsquigarrow \varphi^*$ defines a contravariant functor $C : \mathfrak{M} \longrightarrow \mathfrak{R}$ from the category $\mathfrak{M}$ of manifolds to the category $\mathfrak{R}$ of commutative rings with identity, respectively commutative $\mathbb{R}$-algebras with identity.

Now let X be a G-manifold. According to proposition 1.1.9 and the remark 1.1.10, CX is a G^o-ring, i.e. a ring on which G operates from the right. If $\tau : G \longrightarrow \mathrm{Aut}\, X$ is the given operation, $\tau^* : G \longrightarrow \mathrm{Aut}\, CX$ shall denote the induced operation. We repeat the definition: $\tau_g^* f = f \circ \tau_g$ for $f \in CX$.

Exercise 4.1.1. Let X and X' be manifolds, CX and CX' the corresponding sets of real-valued functions. Show that an arbitrary ring homomorphism $CX' \longrightarrow CX$ is a homomorphism of $\mathbb{R}$-algebras.

Exercise 4.1.2. Let the situation be as in exercise 4.1.1 and $\varphi_i : X \longrightarrow X'$ (i = 1, 2) be maps such that $\varphi_1^* = \varphi_2^*$. Show that then $\varphi_1 = \varphi_2$.

Exercise 4.1.3. Let the situation be as in exercise 4.1.1 and consider the map

$$[X, X'] \longrightarrow [CX', CX]$$

from maps $X \longrightarrow X'$ to ring homomorphisms $CX' \longrightarrow CX$ defined by $\varphi \rightsquigarrow \varphi^*$. Exercise 4.1.2 shows that this map is injective. Show that for paracompact manifolds X, X' this map is bijective. (Hint: Try to imitate the theory of duality for Λ-modules over a ring Λ, considering CX as the dual space of X. The study of the bidual space will then give the desired result.) This result should allow on principle a complete algebraisation of the theory of differentiable manifolds.

Exercise 4.1.4. A manifold X is connected if and only if the ring CX is not decomposable in a direct product of non-trivial rings.

4.2. Operators and vectorfields.

Let X be a manifold and CX the set of real-valued functions, considered as an $\mathbb{R}$-vectorspace.

DEFINITION 4.2.1. An operator A on X is an $\mathbb{R}$-linear map $A : CX \longrightarrow CX$.

Example 4.2.2. An automorphism of CX is an operator. A vectorfield on X is an operator. More generally, a differential operator on X is an operator.

Let OX denote the $\mathbb{R}$-algebra of operators on X. If X' is another manifold and $\varphi : X \longrightarrow X'$ a diffeomorphism, then φ induces an isomorphism $\varphi_* : OX \longrightarrow OX'$ by the definition $\varphi_* A = \varphi^{*-1} \circ A \circ \varphi^*$. This definition means that the following diagram commutes

$$\begin{array}{ccc} CX & \xleftarrow{\ \varphi^*\ } & CX' \\ \uparrow{\scriptstyle A} & & \uparrow{\scriptstyle \varphi_* A} \\ CX & \xleftarrow{\ \varphi^*\ } & CX' \end{array}$$

It is clear that the correspondence $X \rightsquigarrow OX$, $\varphi \rightsquigarrow \varphi_*$ defines a covariant functor $O : \mathfrak{M}_{iso} \longrightarrow \mathfrak{A}_{iso}$ from the category of manifolds and diffeomorphisms to the category of $\mathbb{R}$-algebras and algebra isomorphisms.

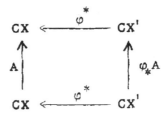

Now let X be a G-manifold with respect to a homomorphism $\tau : G \longrightarrow \operatorname{Aut} X$. Then according to proposition 1.1.9, OX is a G-object in the category of $\mathbb{R}$-algebras. Moreover, the invariant elements under this operation form an $\mathbb{R}$-subalgebra of OX, as follows immediately.

Let us consider an arbitrary associative Λ-algebra O over a ring Λ with identity. Then one can define a new multiplication $[\ ,\ ] : O \times O \longrightarrow O$ in the following way:

$$[A_1, A_2] = A_1A_2 - A_2A_1 \qquad \text{for} \quad A_1, A_2 \in O$$

This multiplication is bilinear and satisfies

1) $[A, A] = O$ for $A \in O$

2) $[A_1, [A_2, A_3]] + [A_2, [A_3, A_1]] + [A_3, [A_1, A_2]] = O$

for $A_1, A_2, A_3 \in O$ (Jacobian identity)

turning therefore O into a Lie-algebra according to

DEFINITION 4. 2. 3. A Λ-module O over a ring Λ with a bilinear map $[\ ,\ ] : O \times O \longrightarrow O$ satisfying $[A, A] = O$ for $A \in O$ and the Jacobian identity is a Lie-algebra over Λ.

DEFINITION 4. 2. 3[1]. A homomorphism $h: O \longrightarrow O'$ of Lie algebras O and O' over a field Λ is a Λ-linear map satisfying

$$h[A_1, A_2] = [hA_1, hA_2] \qquad \text{for} \quad A_1, A_2 \in O.$$

Starting from an associative Λ-algebra O we have associated to O a Λ-Lie algebra. This construction is functorial, i.e. if $h: O \longrightarrow O'$ is a homomorphism of Λ-algebra, then Λ is also a homomorphism of the associated Λ-Lie algebra. Applying this to the $\mathbb{R}$-algebra of operators on X, we obtain

PROPOSITION 4.2.4. Let X be a G-manifold and OX the set of operators on X. The definition $(\tau_g)_*(A) = \tau_g^{*-1} \circ A \circ \tau_g^*$ for $A \in OX$ makes OX a G-set. This operation conserves the $\mathbb{R}$-algebra structure on OX as well as the associated structure of an $\mathbb{R}$-Lie algebra. In particular, the invariant elements under this operation form a $\mathbb{R}$-algebra and a $\mathbb{R}$-Lie algebra respectively.

PROPOSITION 4.2.5. Let X be a G-manifold, X' a G'-manifold and $\varphi : X \longrightarrow X'$ a ρ-equivariant diffeomorphism with respect to a homomorphism $\rho : G \longrightarrow G'$. Then $\varphi_* : OX \longrightarrow OX'$ is a ρ-equivariance with respect to the operations defined in proposition 4.2.4. Moreover, φ_* sends G-invariant operators on X into $\rho(G)$-invariant operators on X'.

This follows from remark 1.1.10 and propositions 1.2.9 and 1.2.12.

We now apply this to vectorfields. Let X be a manifold and A a vectorfield on X. Then A is a map $A : CX \longrightarrow CX$ which satisfies

(i) $A(f_1 + f_2) = Af_1 + Af_2$ for $f_1, f_2 \in CX$

(ii) $A(f_1 f_2) = Af_1 \cdot f_2 + f_1 \cdot Af_2$ for $f_1, f_2 \in CX$

(iii) $A(\lambda) = 0$ for $\lambda \in \mathbb{R}$

Therefore $A \in OX$. In fact, these properties are characteristic for vectorfields. The composition of vectorfields in OX is not a vectorfield, but the composition of vectorfields with respect to the associated $\mathbb{R}$-Lie algebra structure $[\ ,\ ] : OX \times OX \longrightarrow OX$ gives a vectorfield. Here (ii) is essential.

Thus the vectorfields form a subalgebra of this $\mathbb{R}$-Lie algebra. Let DX denote the $\mathbb{R}$-Lie algebra of all vectorfields on X.

If X and X' are manifolds and $\varphi: X \longrightarrow X'$ a diffeomorphism, then the isomorphism $\varphi_* : OX \longrightarrow OX'$ defined at the beginning of this section certainly sends DX into DX'. Applying proposition 4.2.4 we therefore obtain

COROLLARY 4.2.6. Let X be a G-manifold and DX the $\mathbb{R}$-Lie algebra of vectorfields on X. The definition $(\tau_g)_*(A) = \tau_g^{*-1} \circ A \circ \tau_g^*$ for $A \in DX$ makes DX a G-Lie algebra with respect to $\tau : G \longrightarrow \text{Aut } DX$. In particular, the invariant elements of DX under this operation form a $\mathbb{R}$-Lie algebra.

And proposition 4.2.5 gives

COROLLARY 4.2.7. Let X be a G-manifold, X' a G'-manifold and $\varphi : X \longrightarrow X'$ a ρ-equivariant diffeomorphism with respect to a homomorphism $\rho : G \longrightarrow G'$. Then $\varphi_* : DX \longrightarrow DX'$ is a ρ-equivariance with respect to the operations defined in corollary 4.2.6. Moreover, φ_* sends G-invariant vectorfields on X into $\rho(G)$-invariant vectorfields on X'.

For later use, we make explicit the effect of φ_*.

LEMMA 4.2.8. Let $\varphi: X \longrightarrow X'$ be a diffeomorphism and $\varphi_* : DX \longrightarrow DX'$ the induced isomorphism on vectorfields, defined by $\varphi_* A = \varphi^{*-1} \circ A \circ \varphi^*$. Let $x \in X$ and $f' \in CX'$. Then $(\varphi_* A)_{\varphi(x)} f' = A_x(\varphi^* f')$. If $\varphi_{*_x} : T_x(X) \longrightarrow T_{\varphi(x)}(X')$ denotes the linear map of tangent spaces induced by φ, then $(\varphi_* A)_{\varphi(x)} = \varphi_{*_x} A_x$.

Proof: $((\varphi_* A)f')(\varphi(x)) = \varphi^*((\varphi_* A)f')(x) = ((A\varphi^*)f')(x)$ by definition of φ_*. This means $(\varphi_* A)_{\varphi(x)} f' = A_x(\varphi^* f')$. The right side is exactly the definition of $(\varphi_{*_x} A_x)f'$ and therefore also $(\varphi_* A)_{\varphi(x)} = \varphi_{*_x} A_x$.

4. 3. The Lie algebra of a Lie group.

Let G be a Lie group.

DEFINITION 4. 3. 1. The Lie algebra LG of G is the $\mathbb{R}$-Lie algebra of invariant vectorfields under the operation of G on G by left-translations.

Explicitly stated, this means that $A \in LG$ if and only if $(L_g)_* A = A$ for all $g \in G$. LG is a Lie algebra by corollary 4. 2. 6. The letter L shall remind us of left invariant as well as the founder of the theory, Sophus Lie.

The following lemma insures the existence of many left invariant vectorfields on a Lie group.

LEMMA 4. 3. 2. Let G be a Lie group, LG its Lie algebra, G_e the tangent space of G at the identity e and $A_e \in G_e$. Then there exists one and only one $\widetilde{A} \in LG$ such that $\widetilde{A}_e = A_e$.

Proof: If $\widetilde{A}$ exists, then $\widetilde{A} = (L_g)_* \widetilde{A}$ for $g \in G$ and in particular $\widetilde{A}_g = ((L_g)_* \widetilde{A})_g$. In view of lemma 4. 2. 8 this means

$$(1) \quad \widetilde{A}_g = (L_g)_{*_e} A_e .$$

This necessary condition for $\widetilde{A}$ shows the uniqueness. We now define $\widetilde{A}$ by this formula. As $L_e = 1_G$, we certainly have $\widetilde{A}_e = A_e$. The left invariance of $\widetilde{A}$ is seen from

$$((L_g)_* \widetilde{A})_{g\gamma} = (L_g)_{*_\gamma} \widetilde{A}_\gamma = (L_g)_{*_\gamma} (L_\gamma)_{*_e} A_e$$

$$= (L_{g\gamma})_{*_e} A_e = \widetilde{A}_{g\gamma}.$$

There remains to show that the family $(\widetilde{A}_g)_{g\in G}$ is a vectorfield (i.e. a differentiable vectorfield), which means that $\widetilde{A}(CG) \subset CG$. Let $f \in CG$. By lemma 4.2.8

$$((L_g)_* \widetilde{A})_g f = A_e(L_g^* f)$$

and therefore

$$(\widetilde{A} f)(g) = A_e(L_g^* f).$$

Let $\gamma : I \longrightarrow G$, I an interval of $\mathbb{R}$ containing 0, a differentiable curve in G with $\frac{d}{dt} \gamma_t \Big/_{t=o} = A_e$. Then

$$A_e(L_g^* f) = \frac{\partial}{\partial t} \left[(L_g^* f)(\gamma_t)\right]_{t=o} = \frac{\partial}{\partial t} f(g\gamma_t)\Big|_{t=o}$$

which shows $\widetilde{A} f \in CG$. ∎

The correspondence $A_e \rightsquigarrow \widetilde{A}$ of the lemma defines a bijective map

$$\aleph : G_e \longrightarrow LG$$

which is seen to be an isomorphism of $\mathbb{R}$-vectorspaces by (1). We have proved

THEOREM 4.3.3. <u>Let</u> G <u>be a Lie group,</u> LG <u>its Lie algebra and</u> G_e <u>the tangent space of</u> G <u>at the identity</u> e. <u>The formula</u>

$$(\aleph(A_e))_g = (L_g)_{*_e} A_e \qquad \underline{\text{for}} \quad g \in G,\ A_e \in G_e$$

<u>defines a map</u> $\aleph : G_e \longrightarrow LG$, <u>which is an isomorphism of</u> $\mathbb{R}$<u>-vectorspaces.</u>

This map $\aleph$ allows transporting the structure of $\mathbb{R}$-Lie algebra from LG to G_e. In this sense, G_e is often referred to as the Lie algebra of G.

COROLLARY 4.3.4. <u>Let</u> G <u>be a Lie group of dimension</u> n. <u>The Lie algebra</u> LG <u>is a Lie algebra of dimension</u> n.

Consider the map $(L_g)_{*_e} : G_e \longrightarrow G_g$, which is an isomorphism for all $g \in G$. More generally, the maps $P(g_1, g_2) = (L_{g_2})_{*_e} (L_{g_1})_{*_e}^{-1} : G_{g_1} \longrightarrow G_{g_2}$ have the properties:

1) $P(g_2, g_3)\, P(g_1, g_2) = P(g_1, g_3)$ for $g_1, g_2, g_3 \in G$

2) $P(g, g) = 1_{G_g}$ for $g \in G$

DEFINITION 4.3.5. Let X be a manifold and $P(g_2, g_1) : T_{g_1}(X) \longrightarrow T_{g_2}(X)$ an $\mathbb{R}$-linear map for all $(g_1, g_2) \in X \times X$, satisfying 1) and 2). Then X is called a <u>parallelizable manifold.</u>

The maps $P(g_2, g_1)$ are then necessarily isomorphisms and it makes sense to speak of the dimension of X.

Let e be a fixed point of X and A_{i_e} $(i = 1, \cdots, n,\ n = \dim X)$ a base of the vectorspace $T_e X$. Then $P(e, g) A_{i_e} = A_{i_g}$ defines vectorfields A_i $(i = 1, \cdots, n)$ on X such that the vectors A_{i_g} $(i = 1, \cdots, n)$ form a base of $T_g X$ for all $g \in G$.

COROLLARY 4.3.6. The manifold of a Lie group G is parallelizable.

Example 4.3.7. Consider $\mathbb{R}$ with its additive Lie group structure. Then $L\mathbb{R} \cong \mathbb{R}$ as vectorspace, because the tangent space of $\mathbb{R}$ at O is $\mathbb{R}$. There is only one possible Lie algebra structure on $\mathbb{R}$, defined by $[\lambda_1, \lambda_2] = O$ for $\lambda_1, \lambda_2 \in \mathbb{R}$.

By the same argument, $L\mathbb{T} = \mathbb{R}$ for the additive group $\mathbb{T} = \mathbb{R}/\mathbb{Z}$.

Now let V be n-dimensional $\mathbb{R}$-vectorspace and $G = GL(V)$. We first remark that considering $GL(V) \subset \mathcal{L}(V)$ = algebra of $\mathbb{R}$-linear endomorphiams of V, the tangent space G_g is identified to $\mathcal{L}(V)$ for all $g \in G$. The multiplication in $GL(V)$ is the restriction of the bilinear map $\mathcal{L}(V) \times \mathcal{L}(V) \longrightarrow \mathcal{L}(V)$ defining the multiplication in $\mathcal{L}(V)$. This shows that $(L_g)_{*_\gamma} A_\gamma = gA_\gamma$ for $g \in GL(V)$, $A_\gamma \in G_\gamma$ identified to $\mathcal{L}(V)$. We show now

PROPOSITION 4.3.8. After the canonical identification of $L(GL(V))$ with the tangent space at the identity, we have $L(GL(V)) = \mathcal{L}(V)$ as Lie algebras, where on $\mathcal{L}(V)$ we consider the Lie algebra structure associated in the sense of section 4.2 to the natural algebra structure.

Proof: Let $A_1, A_2 \in L(GL(V))$. We use the formula

$$[A_1, A_2]_g = \left(\frac{d}{dg} A_{2_g}\right)(g) A_{1_g} - \left(\frac{d}{dg} A_{1_g}\right)(g) A_{2_g}$$

which is valid for the global chart given by the embedding $GL(V) \subset \mathcal{L}(V)$.

In view of $A_{i_g} = g A_{i_e}$ we have

$$\left(\frac{d}{dg} A_{i_g}\right)(g) A_{j_g} = A_{j_g} A_{i_g}$$

which shows

$$[A_1, A_2]_g = A_{1_g} A_{2_g} - A_{2_g} A_{1_g} .$$

But the right side is just the commutator $[A_{1_g}, A_{2_g}]$ in $\mathcal{L}(V)$. For $g = e$ this gives the desired result.

We have defined the Lie algebra LG of a Lie group G by consideration of the operation of G on G by left translation. Doing the same for the right translations, we obtain another Lie algebra RG. Explicitly: RG is the Lie algebra of the right invariant vectorfields. As in theorem 4.3.3, we can define an isomorphism $G_e \longrightarrow RG$ of $\mathbb{R}$-vectorspaces, obtaining therefore an isomorphism $LG \cong RG$ of $\mathbb{R}$-vectorspaces. We shall see in section 4.6 that there is also a natural isomorphism $LG \cong RG$ of the $\mathbb{R}$-Lie algebra structure.

Exercise 4.3.9. Let G be a Lie group, CG the $\mathbb{R}$-vectorspace of real-valued functions on G, DG the $\mathbb{R}$-Lie algebra of all vectorfields on G and LG the Lie algebra of G. Show that $DG = CG \otimes_{\mathbb{R}} LG$.

4.4. Effect of maps on operators and vectorfields.

In section 4.2 we have seen the effect of diffeomorphisms on operators. We want to study now the effect of arbitrary (i.e. differentiable) maps.

Let X, X' be manifolds and A, A' operators on X, X' respectively.

DEFINITION 4.4.1. A and A' are *φ-related* with respect to a map $\varphi: X \longrightarrow X'$, if the following diagram commutes.

$$\begin{array}{ccc} CX & \overset{\varphi^*}{\text{———}} & CX' \\ A \uparrow & & \uparrow A' \\ CX & \overset{\varphi^*}{\text{———}} & CX' \end{array}$$

If φ is a diffeomorphism, A and $\varphi_* A$ are φ-related operators. But in the general case, A does neither determine an A' such that A and A' are φ-related, nor is A' unique, if it exists.

LEMMA 4.4.2. *Let $\varphi: X \longrightarrow X'$ be a map.*

(i) *If A_i and A_i' $(i = 1, 2)$ are φ-related operators on X and X' respectively, then the following operators are φ-related:*

$$A_1 + A_2 \quad \text{and} \quad A_1' + A_2' \,,$$

$$A_1 A_2 \quad \text{and} \quad A_1' A_2' \,,$$

$$[A_1, A_2] \quad \text{and} \quad [A_1', A_2'] \,.$$

(ii) *If A and A' are φ-related operators on X and X' respectively, then for $\lambda \in \mathbb{R}$, λA and $\lambda A'$ are φ-related.*

Proof: (i) Let $f' \in CX'$. Then

$$\varphi^*((A_1' + A_2')f') = \varphi^*(A_1' f' + A_2' f') = \varphi^*(A_1' f') + \varphi^*(A_2' f')$$

$$= A_1(\varphi^* f') + A_2(\varphi^* f') = (A_1 + A_2)(\varphi^* f'),$$

showing that $A_1 + A_2$ and $A_1' + A_2'$ are φ-related. The φ-relatedness of A_1A_2 and $A_1'A_2'$ is seen by comparing the diagrams serving to define φ-relatedness, and the third assertion is a consequence of this and (ii).

$$\text{(ii)}\quad \varphi^*((\lambda A')f') = \varphi^*(\lambda(A'f')) = \varphi^*\lambda \cdot \varphi^*(A'f')$$

$$= \lambda \cdot A(\varphi^* f') = (\lambda A)(\varphi^* f') \quad , \text{ q. e. d.}$$

The lemma applies in particular to vectorfields. For that we make explicit the notion of φ-relatedness in

PROPOSITION 4. 4. 3. _Let_ X, X' _be manifolds_, $\varphi : X \longrightarrow X'$ _a map and_ A, A' _vectorfields on_ X, X' _respectively. Then_ A _and_ A' _are_ φ _-related if and only if_ $\varphi_{*_x} A_x = A'_{\varphi(x)}$ _for every_ $x \in X$.

Proof: Let $f' \in CX'$. Then

$$(\varphi_{*_x} A_x) f' = A_x(\varphi^* f') = (A(\varphi^* f'))(x)$$

by definition of φ_{*_x}. On the other hand

$$A'_{\varphi(x)} f' = (A'f')(\varphi(x)) = (\varphi^*(A'f'))(x) .$$

Comparison proves the lemma.

4. 5. The functor L.

We have defined the Lie algebra LG for any Lie group G. We want to extend this correspondence to a functor from Lie groups to Lie algebras.

LEMMA 4. 5. 1. _Let_ G, G' _be Lie groups_, $\rho : G \longrightarrow G'$ _a homomorphism of Lie groups and_ $A \in LG$. _Then there exists one and only one_ $A' \in LG'$, _such that_ A _and_ A' _are_ φ_-related._

Proof: Suppose A' exists with the desired properties. By proposition 4.4.3 we obtain

$$(1) \qquad A'_{e'} = \rho_{*_e} A_e ,$$

where e, e is the identity of G, G' respectively. By lemma 4.3.2 there is only one $\widetilde{A}' \in LG'$ such that $\widetilde{A}'_{e'} = A'_{e'}$. This proves uniqueness. Now we define conversely A' as the unique element of LG' satisfying (1). There remains to show that A and A' are φ-related. Now $\rho \circ L_g = L_{\rho(g)} \circ \rho$ implies

$$\rho_{*_g} A_g = \rho_{*_g} (L_g)_{*_e} A_e = (L_{\rho(g)})_{*_{e'}} \rho_{*_e} A_e = A'_{\rho(g)}$$

which is the desired result in view of proposition 4.4.3.

In the proof of the lemma 4.5.1, we used only the restriction of ρ to a neighborhood of $e \in G$. It is useful to introduce a corresponding notion.

DEFINITION 4.5.2. Let G, G' be Lie groups and U an open neighborhood of $e \in G$. A local homomorphism $G \longrightarrow G'$ defined on U is a differentiable map $\rho : U \longrightarrow G'$ which satisfies

$$\rho(g_1 g_2) = \rho(g_1)\rho(g_2) \text{ for all } g_1, g_2 \in U \text{ such that } g_1 g_2 \in U .$$

The restriction of a homomorphism $\rho : G \longrightarrow G'$ to an open neighborhood of $e \in G$ is a local homomorphism $G \longrightarrow G'$. If we identify a map with its restrictions to open subsets of the domain, we can compose local homomorphisms, obtaining thus the category of Lie groups and local homomorphisms. An equivalence in this category is called a local isomorphism. Explicitly stated we have

DEFINITION 4.5.3. Two Lie groups, G and G', are *locally isomorphic* if and only if there exists open neighborhoods U, U' of e, e' and a diffeomorphism $\rho: U \longrightarrow U'$ with inverse $\rho': U' \longrightarrow U$ such that both ρ and ρ' are local homomorphisms.

THEOREM 4.5.4. *Let* G, G' *be Lie groups,* U *an open neighborhood of* e *in* G *and* $\rho: U \longrightarrow G'$ *a local homomorphism. The formula*

$$(L(\rho)A)_{e'} = \rho_{*_e} A_e \quad \text{for } A \in LG$$

defines a homomorphism of Lie algebras $L(\rho): LG \longrightarrow LG'$. *The following diagram is commutative,*

$$\begin{array}{ccc} G_e & \xrightarrow{\rho_{*_e}} & G'_{e'} \\ \aleph \downarrow & & \downarrow \aleph \\ LG & \xrightarrow{L(\rho)} & LG' \end{array}$$

where $\aleph$ *denotes the isomorphism of theorem 4.3.3. Moreover for* $A \in LG$ *the vectorfields* A/U *and* $L(\rho)A \in LG'$ *are* ρ*-related.*

Proof: In the proof of lemma 4.5.1 all was shown except the fact that $L(\rho)$ is a homomorphism. This is a consequence of lemma 4.4.2.

We observe that a homomorphism $\rho: G \longrightarrow G'$ defines in the same way a homomorphism $R(\rho): RG \longrightarrow RG'$ of the Lie algebras of right invariant vectorfields.

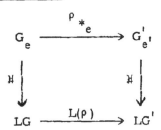

Complement to 4.5.4. If $\rho : G \longrightarrow G'$ is an isomorphism, then $L(\rho) = \rho_*/LG$, where $\rho_* : DG \longrightarrow DG'$ is the map defined in section 4.2.

Proof: We have to show the commutativity of the diagram

$$\begin{array}{ccc} LG & \xrightarrow{L(\rho)} & LG' \\ \cap & & \cap \\ DG & \xrightarrow{\rho *} & DG' \end{array}$$

Let $A \in LG$. Then on one hand

$$(L(\rho)A)_{\rho(g)} = (L_{\rho(g)})_{*_e} \rho_{*_e} A_e = \rho_*(L_g)_{*_e} A_e$$

and on the other hand

$$(\rho_* A)_{\rho(g)} = \rho_{*_g} A_g$$

by lemma 4.2.8. This shows the desired property.

One cannot define $L(\rho)$ in general by ρ_*, because this map only makes sense for a diffeomorphism ρ.

THEOREM 4.5.5. Let $\mathfrak{Lg}$ be the category of Lie groups and local homomorphisms of Lie groups, $\mathfrak{LA}$ the category of $\mathbb{R}$-Lie algebras and homomorphisms of Lie algebras. The correspondence $G \rightsquigarrow LG$, $\rho \rightsquigarrow L(\rho)$ defines a covariant functor $L : \mathfrak{Lg} \longrightarrow \mathfrak{LA}$.

This is clear by theorem 4.5.4.

We can also consider the functor given by $G \rightsquigarrow G_e$, $\rho \rightsquigarrow \rho_{*_e}$. The commutativity of the diagram in theorem 4.5.4 expresses that $\aleph$ is a natural transformation of this functor into L, in fact a natural equivalence.

COROLLARY 4. 5. 6. <u>The Lie algebras of locally isomorphic groups are isomorphic.</u>

<u>Proof:</u> L sends equivalences in $\mathcal{L}\mathfrak{G}$ into equivalences in $\mathcal{L}\mathfrak{A}$. We apply this to the natural injection $G_0 \hookrightarrow G$ of the connected component of the identity G_0 into G, which is a local isomorphism. This shows $LG_0 \cong LG$. The Lie algebra is therefore a property of G_0, in fact, of an arbitrary neighborhood of the identity.

Example 4. 5. 7. The canonical homomorphism $\mathbb{R} \longrightarrow \mathbb{T} = \mathbb{R}/\mathbb{Z}$ is a local isomorphism. Therefore $L\mathbb{R} \cong L\mathbb{T}$, what we already know.

LEMMA 4. 5. 8. <u>If</u> $\rho: \mathbb{T} \longrightarrow \mathbb{R}$ <u>is a homomorphism, then</u> $\rho = 0$.

<u>Proof:</u> $\mathbb{T}$ being compact, $\rho(\mathbb{T})$ is contained in a closed interval I. Suppose $t \in \mathbb{T}$ with $\rho(t) \neq 0$. Then there exists a positive integer n such that $n\rho(t) \notin I$, which is a contradiction.

This proves that there is no homomorphism $\mathbb{T} \longrightarrow \mathbb{R}$ inducing the identity isomorphism $L\mathbb{T} = L\mathbb{R}$. But of course the natural local isomorphism has this property.

Example 4. 5. 9. Let V be an $\mathbb{R}$-vectorspace of finite dimension, and $\tau : G \longrightarrow GL(V)$ a representation of the Lie group G in V. We proved in Proposition 4. 3. 8 that $\mathcal{L}(V)$ is the Lie algebra of $GL(V)$. The map τ can be seen to be differentiable, and induces therefore a homomorphism $L(\tau) : LG \longrightarrow \mathcal{L}(V)$.

DEFINITION 4.5.10. Let Λ be a ring, V a Λ-module and $\mathcal{L}(V)$ the Λ-Lie algebra of Λ-endomorphisms of V. A <u>representation of a Λ-Lie algebra O in</u> V is a homomorphism $\sigma: O \longrightarrow \mathcal{L}(V)$. V is then called an O-module with respect to σ.

Following example 4.5.9, a representation of a Lie group G in a finite-dimensional $\mathbb{R}$-vectorspace V defines a representation of the Lie algebra LG in V.

We now consider the homomorphism det : $GL(V) \longrightarrow \mathbb{R}^*$ into the multiplicative group $\mathbb{R}^*$ of the reals. The Lie algebra of $\mathbb{R}^*$ is $\mathbb{R}$. We prove

PROPOSITION 4.5.11. <u>The homomorphism</u> $\mathcal{L}(V) \longrightarrow \mathbb{R}$ <u>of Lie algebras induced by the homomorphism</u> $GL(V) \longrightarrow \mathbb{R}^*$ <u>is the trace map.</u>

<u>Proof:</u> Let $A \in \mathcal{L}(V)$ and α_t a curve in GL(V) with $\alpha_o = e$, $\dot{\alpha}_o = A$. Then

$$\det{}_{*_e} A = \frac{d}{dt}\{\det \alpha_t\}\Big/_{t=0} .$$

Now for any non-degenerated n-form ω on V (n = dim V) and any n-tuple of vectors $v_1, \cdots, v_n$ of V we have

$$\omega(v_1, \cdots, v_n) \cdot \det \alpha_t = \omega(\alpha_t v_1, \cdots, \alpha_t v_n)$$

and therefore

$$\omega(v_1, \cdots, v_n) \cdot \det_{*_e} A = \frac{d}{dt}\{\omega(\alpha_t v_1, \cdots, \alpha_t v_n)\}/_{t=0}$$

$$= \sum_i \omega(\alpha_t v_1, \cdots, \alpha_t v_{i-1}, \dot{\alpha}_t v_i, \alpha_t v_{i+1}, \cdots, \alpha_t v_n)/_{t=0}$$

$$= \sum_i \omega(v_1, \cdots, Av_i, \cdots, v_n)$$

$$= \omega(v_1, \cdots, v_n) \cdot \operatorname{tr} A$$

showing $\det_{*_e} A = \operatorname{tr} A$. This is the desired result in view of theorem 4.5.4.

COROLLARY 4.5.12. $\operatorname{tr}(AB) = \operatorname{tr}(BA)$ for $A, B \in \mathcal{L}(V)$.

<u>Proof:</u> $\operatorname{tr} : \mathcal{L}(V) \longrightarrow \mathbb{R}$ being a homomorphism of Lie algebras, we have for $A, B \in \mathcal{L}(V)$

$$\operatorname{tr}(AB - BA) = \operatorname{tr}[A, B] = [\operatorname{tr} A, \operatorname{tr} B] = 0,$$

the latter bracket being the trivial one in $\mathbb{R}$.

Example 4.5.13. Let G be a Lie group and

$$0 \longrightarrow G_e \longrightarrow TG \longrightarrow G \longrightarrow e$$

the sequence of example 3.1.7. It induces a sequence of Lie algebra homomorphisms

$$0 \longrightarrow L(G_e) \longrightarrow L(TG) \longrightarrow LG \longrightarrow 0 \quad .$$

Here $LG_e \cong G_e$ (see example 4.6.4 below). See 7.5.6 for the exactness of this sequence. The inclusion $G \longrightarrow TG$ induces a homomorphism

4.6. Applications of the functorality of L.

4.6.1. The Lie algebra of a product group. Let G_1, G_2 be Lie groups and $G_1 \times G_2$ the product group. The canonical projections $p_i : G_1 \times G_2 \longrightarrow G_i$ $(i = 1, 2)$ are homomorphisms of Lie groups and induce Lie algebra homomorphisms $L(p_i) : L(G_1 \times G_2) \longrightarrow LG_i$. Let e_1, e_2 be the identities of G_1, G_2. By theorem 4.5.4, we have the commutative diagram (expressing the naturality of $\mathcal{H}$)

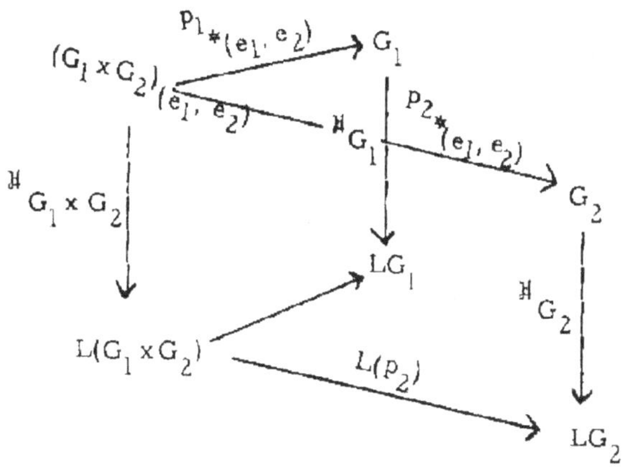

the vertical arrows being isomorphisms. The $\mathbb{R}$-linear isomorphism

$$(G_1 \times G_2)_{e_1, e_2} \cong G_{1_{e_1}} \times G_{2_{e_2}}$$

implies therefore the isomorphisms of $\mathbb{R}$-vectorspaces

$$L(G_1 \times G_2) \cong LG_1 \times LG_2 \; .$$

If $q_i : LG_1 \times LG_2 \longrightarrow LG_i$ $(i = 1, 2)$ denotes the canonical projection, this isomorphism is given by the commutative diagram

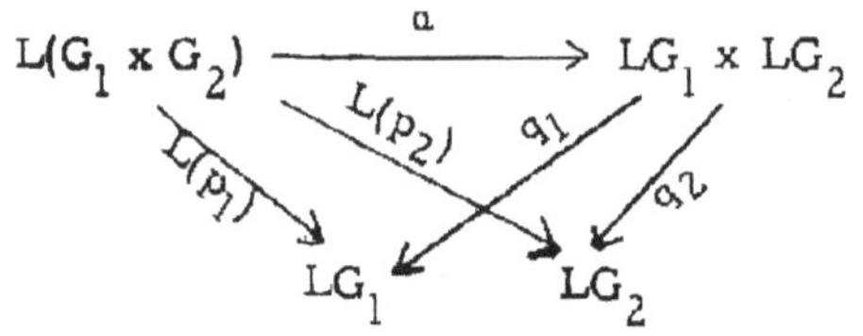

We want to transport the Lie algebra structure of $L(G_1 \times G_2)$ to $LG_1 \times LG_2$. For $A \in L(G_1 \times G_2)$ we have $\alpha(A) = (L(p_1)A,\ L(p_2)A) = (A_1, A_2)$ with $A_i = L(p_i)A$ $(i = 1, 2)$. Similarly for $A' \in L(G_1 \times G_2)$ we have $\alpha(A') = (A_1', A_2')$ with $A_i' = L(p_i)A'$ $(i = 1, 2)$. Then

$$\alpha[A, A'] = (L(p_1)[A, A'] \ , \ L(p_2)[A, A'])$$

$$= ([A_1, A_1'] \ , [A_2, A_2'])$$

as $L(p_i)$ are Lie algebra homomorphisms. We define

$$[\alpha(A), \alpha(A')] = \alpha[A, A']$$

which means

$$(1) \quad [(A_1, A_2), (A_1', A_2')] = ([A_1, A_1'], [A_2, A_2'])$$

for $A_i, A_i' \in LG_i$ $(i = 1, 2)$.

With this definition, α is an isomorphism.

Let more generally Λ be a ring and O_1, O_2 Λ-Lie algebras. Consider the product module $O_1 \times O_2$ with the map $[\ ,\ ]: (O_1 \times O_2) \times (O_1 \times O_2) \longrightarrow O_1 \times O_2$ defined by (1). Then $O_1 \times O_2$ is a Λ-Lie algebra.

DEFINITION 4.6.2. The direct product of two Λ-Lie algebras O_1, O_2 is the Lie algebra $O_1 \times O_2$ with the multiplication defined by (1).

We can now state

PROPOSITION 4.6.3. <u>Let</u> G_1, G_2 <u>be Lie groups and</u> $G_1 \times G_2$ <u>the direct product. The Lie algebra</u> $L(G_1 \times G_2)$ <u>is canonically isomorphic to the direct product</u> $LG_1 \times LG_2$ <u>of the Lie algebras</u> LG_1 <u>and</u> LG_2.

We remark that commutativity for the multiplication in a Lie algebra means that $[A_1, A_2] = 0$ for any pair A_1, A_2. It is then clear that the product of commutative Lie algebras is commutative.

Example 4.6.4. $L(\mathbb{R}^n) = L\mathbb{R} \times \cdots \times L\mathbb{R}$. But we have already seen that $L\mathbb{R} = \mathbb{R}$ with the trivial Lie algebra structure (example 4.3.7). Therefore $L(\mathbb{R}^n) = \mathbb{R}^n$ with the trivial Lie algebra structure.

Similarly $L(\mathbb{T}^n) = \mathbb{R}^n$ for the additive group $\mathbb{T}^n = \mathbb{R}^n/\mathbb{Z}^n$.

4.6.5. The relation between LG and RG. Let G be a Lie group, G^o the opposite group and $I : G \longrightarrow G^o$ the isomorphism defined by $I(g) = g^{-1}$ for $g \in G$.

LEMMA 4.6.6. $I_{*_e} = -1_{G_e}$.

<u>Proof:</u> Consider the map $\varphi: G \longrightarrow G$ defined by $\varphi(g) = gg^{-1}$. φ being constant, $\varphi_{*_g} = 0 : G_g \longrightarrow G_e$. But

$$\varphi_{*_g} = (R_{g^{-1}})_{*_g} + (L_g)_{*_{g^{-1}}} I_{*_g}$$

and therefore

$$I_{*_g} = -(L_g)^{-1}_{*_{g^{-1}}} \circ (R_{g^{-1}})_{*_g} = -(L_{g^{-1}})_{*_e} \circ (R_{g^{-1}})_{*_g}$$

For $g = e$ we obtain

$$I_{*_e} = -1_{G_e}, \qquad \text{q.e.d.}$$

Remark. We have shown the formula

$$I_{*_g} = -(L_{g^{-1}})_{*_e} \circ (R_{g^{-1}})_{*_g} \qquad \text{for } g \in G$$

This means that the tangent map to the map $I : G \longrightarrow G$ in each point is already given by the tangent maps of the translations. This can be used to prove that the differentiability of the multiplication $G \times G \longrightarrow G$ on a group manifold already implies the differentiability of $I : G \longrightarrow G$.

Let Λ be a ring and $\mathfrak{g}$ a Lie algebra over Λ. The _opposite Lie algebra_ $\mathfrak{g}^o$ is the Λ-module $\mathfrak{g}$ with the Lie algebra structure defined by the bracket $[A_1, A_2]^o = [A_1, A_2]$ for $A_1, A_2 \in \mathfrak{g}$.

PROPOSITION 4.6.7. _Let_ G _be a Lie group and_ G^o _the opposite group. Identify_ LG _with_ G_e _and_ $L(G^o)$ _with_ G^o_e _by the canonical isomorphism_ $\aleph$ _of theorem_ 4.3.3. _Then_

$$L(G^o) = (LG)^o$$

$(LG)^o$ _being the opposite Lie algebra of_ LG.

Proof: After the indicated identification we have $L(I) = I_{*_e}$ for the isomorphism $I : G \longrightarrow G^o$. To $A_i \in LG$ $(i = 1, 2)$ corresponds $I_{*_e} A_i = -A_i \in L(G^o)$. To $[A_1, A_2] \in LG$ there corresponds on one hand $-[A_1, A_2]_{LG}$ and on the other hand also $[-A_1, -A_2]_{L(G^o)}$, as I is an isomorphism. Therefore

$$[A_1, A_2]_{L(G^o)} = -[A_1, A_2]_{LG}$$

$$= [A_1, A_2]_{(LG)^o}, \qquad \text{q. e. d.}$$

COROLLARY 4.6.8. *Let G be a Lie group, LG the Lie algebra of left invariant vectorfields and RG the Lie algebra of right invariant vectorfields. Identify LG and RG with G_e by the canonical isomorphism of theorem 4.3.3. Then*

$$RG = (LG)^{o}$$

Proof: We observe that left translations of G are right translations of G^o and vice versa, so that $LG = R(G^o)$ and $RG = L(G^o)$. After the mentioned identifications, by proposition 4.6.7 $L(G^o) = (LG)^o$, which shows $RG = (LG)^o$.

This shows, of course, the existence of a natural isomorphism $RG \cong LG$. Moreover

COROLLARY 4.6.9. *Let G be a Lie group. If G is commutative, then LG is commutative.*

Proof: The commutativity of G implies $RG = LG$. But $RG = (LG)^o$ and therefore $LG = (LG)^o$, q.e.d.

We shall see in chapter 6 that, for connected G, the converse is also true.

Example 4.6.10. Let V be a finite-dimensional $\mathbb{R}$-vectorspace and consider the natural representation of GL(V) in V. We have seen that identifying L(GL(V)) canonically with the tangent space at the identity, we obtain $\mathcal{L}(V)$. By corollary 4.6.8, RG identified with the tangent space at the identity is $(\mathcal{L}(V))^o$.

4.7. The adjoint representation of a Lie group.

Consider the operation of G on G by inner automorphisms $\mathfrak{J} : G \longrightarrow \operatorname{Aut} G$ (see example 1.1.5). The functor L transforms the G-group G into a G-Lie algebra LG according to proposition 1.1.9. We repeat the definition of the induced G-operation on LG: it is the composed homomorphism

$$G \xrightarrow{\mathfrak{J}} \operatorname{Aut} G \xrightarrow{L} \operatorname{Aut} LG$$

DEFINITION 4.7.1. The *adjoint representation of a Lie group* G is the representation of G in LG induced by the operation of G on G by inner automorphisms: $\operatorname{Ad} g = L(\mathfrak{J}_g)$.

PROPOSITION 4.7.2. *Let* G, G' *be Lie groups and* $\rho : G \longrightarrow G'$ *a homomorphism. Then* $L(\rho) : LG \longrightarrow LG'$ *is a* ρ*-equivariance with respect to the adjoint representations of* G, G' *in* LG, LG'.

Proof: The commutativity of the diagram

$$\begin{array}{ccc} G & \xrightarrow{\rho} & G' \\ {\scriptstyle \mathfrak{J}_g}\downarrow & & \downarrow{\scriptstyle \mathfrak{J}_{\rho(g)}} \\ G & \xrightarrow{\rho} & G' \end{array}$$

(see also example 1.2.3) and the functorality of L prove the statement.

Consider the canonical isomorphism $\aleph : G_e \longrightarrow LG$ of theorem 4.3.3 permitting to interpret G_e as the Lie algebra of G. From theorem 4.5.4 it follows that the effect of $\operatorname{Ad} g : LG \longrightarrow LG$ on G_e is given by the map

$(\mathfrak{J}_g)_{*_e} : G_e \longrightarrow G_e$. This proves that the operation of G on G_e defined in example 3.2.8 is just the adjoint representation after identification of G_e and LG.

Another description of the adjoint representation is given in

PROPOSITION 4.7.3. Let G be a Lie group, $Ad : G \longrightarrow Aut\ LG$ the adjoint representation and $A \in LG$. Then

$$Ad\,g\,A = (R_{g^{-1}})_* A$$

Proof: $Ad\,g\,A = L(\mathfrak{J}_g)A = (\mathfrak{J}_g)_* A$ by complement 4.5.4. But $(\mathfrak{J}_g)_* A = (R_{g^{-1}})_*(L_g)_* A = (R_{g^{-1}})_* A$ for $A \in LG$.

This shows that the operation of G in G by right translations defines a right operation of G in LG and the adjoint representation describes the effect of this operation.

Chapter 5. VECTORFIELDS AND 1-PARAMETER GROUPS OF TRANSFORMATIONS.

The Lie algebra of a Lie group gives a deep information on the group. The key for the understanding of this is the relation between vectorfields and ordinary differential equations, which is studied in this chapter.

5.1. 1-parameter groups of transformations.

DEFINITION 5.1.1. An $\mathbb{R}$-manifold X is called a 1-parameter group of transformations.

Let X be a manifold and $\gamma: I \longrightarrow X$, $t \rightsquigarrow \gamma_t$, a curve on X. Here I denotes an open interval of $\mathbb{R}$ containing 0. We denote by $\dot{\gamma}_t = \frac{d}{dt}\gamma_t$ the tangent vector of γ in the point γ_t. We then have

$$\dot{\gamma}_t f = \frac{d}{dt} f(\gamma_t) \qquad \text{for every } f \in CX$$

which characterizes $\dot{\gamma}_t$.

Now let $\varphi: \mathbb{R} \times X \longrightarrow X$, $(t, x) \rightsquigarrow \varphi_t(x)$, be a 1-parameter group of transformations. We shall also say for this situation: "φ is a 1-parameter group of transformations of X." Define for every $x \in X$

$$(1) \qquad A_x = \frac{d}{dt}\varphi_t(x)\Big/_{t=0} = \dot{\varphi}_t(x)\Big/_{t=0}$$

Then $A = (A_x)_{x \in X}$ is a vectorfield on X.

If $t \rightsquigarrow \varphi_t$ is a 1-parameter group of transformations, then also $t \rightsquigarrow \varphi_{st} = \Psi_t$. If A, B are the vectorfields induced by φ_t, Ψ_t respectively, then

$$B_x = \frac{d}{dt}\{\Psi_t(x)\}_{t=0} = \frac{d}{dt}\varphi_{st}(x)\Big/_{t=0} = s A_x .$$

In this definition of the vectorfield corresponding to a 1-parameter group of transformations, we have not made use of the fact that φ_t is globally defined.

DEFINITION 5.1.2. Let $\epsilon > 0$, I_ϵ be an open interval $(-\epsilon, \epsilon)$ of $\mathbb{R}$ and U an open set of X. A <u>local 1-parameter group of local transformations</u> of X defined on U is a map $\varphi : I_\epsilon \times U \longrightarrow X$, $(t, x) \rightsquigarrow \varphi_t(x)$ such that

1) for all $t \in I_\epsilon$, $\varphi_t : U \longrightarrow \varphi_t(U)$ is a diffeomorphism,

2) if $t, s, t+s \in I_\epsilon$ and $x, \varphi_s(x) \in U$, then $\varphi_{t+s}(x) = \varphi_t(\varphi_s(x))$.

Equation (1) still makes sense for $x \in U$ and defines a vectorfield A on U, which is called the vectorfield induced on U by φ_t. The properties of a local 1-parameter group of local transformations are not used for this fact, but they allow to prove

PROPOSITION 5.1.3. <u>Let</u> $\varphi : I_\epsilon \times U \longrightarrow X$ <u>be a local 1-parameter group of transformations of</u> X, <u>where</u> $\epsilon > 0$, $I_\epsilon = (-\epsilon, \epsilon) \subset \mathbb{R}$, U <u>an open subset of</u> X, <u>and</u> A <u>the induced vectorfield on</u> U. <u>Then for</u> $x \in U$, <u>the curve</u> $t \rightsquigarrow \varphi_t(x)$ <u>satisfies the differential equation</u>

$$\dot{\varphi}_t(x) = A_{\varphi_t(x)}$$

with the initial condition

$$\varphi_0(x) = x .$$

Proof: Let $f \in CX$. Then for fixed $x \in U$

$$\dot{\varphi}_t(x)f = \frac{d}{dt} f(\varphi_t(x)) = \lim_{s \to 0} \frac{1}{s} \{ f(\varphi_{t+s}(x)) - f(\varphi_t(x)) \}$$

$$= \lim_{s \to 0} \frac{1}{s} \{ f(\varphi_s(\varphi_t(x))) - f(\varphi_t(x)) \}$$

$$= \frac{d}{ds} f(\varphi_s(\varphi_t(x))) \Big/_{s=0}$$

$$= A_{\varphi_t(x)} f \,, \qquad \text{q. e. d.}$$

COROLLARY 5.1.4. Let $\varphi, \Psi : I_\epsilon \times U \longrightarrow X$ be two local 1-parameter groups of local transformations defined on U. If they induce the same vectorfields on U, they coincide.

Proof: This is just the uniqueness theorem for ordinary differential equations.

Applying the existence theorem for ordinary differential equations, we prove now

PROPOSITION 5.1.5. Let A be a vectorfield on X. For any $x \in X$ there exists an $\epsilon > 0$, an open subset U of X and a local 1-parameter group of local transformations $\varphi : I_\epsilon \times U \longrightarrow X$, which induces on U the given vectorfield.

Proof: For a fixed $x \in X$ we define $t \rightsquigarrow \varphi_t(x)$ as the solution of the differential equation

$$\dot{\varphi}_t(x) = A_{\varphi_t(x)}$$

with initial value $\varphi_o(x) = x$. We now prove

$$\varphi_t(\varphi_s(x)) = \varphi_{t+s}(x) \qquad \text{for} \quad s, t, t+s \in I_\epsilon$$

such that both sides are defined. Write $\varphi_{t+s}(x) = \alpha_1(t), \varphi_t(\varphi_s(x)) = \alpha_2(t)$.

Then

$$\dot{\alpha}_1(t) = \dot{\varphi}_{t+s}(x) = A_{\varphi_{t+s}(x)} = A_{\alpha_1(t)}$$

and

$$\dot{\alpha}_2(t) = \dot{\varphi}_t(\varphi_s(x)) = A_{\varphi_t(\varphi_s(x))} = A_{\alpha_2(t)} .$$

α_i $(i = 1, 2)$ are solutions of the same differential equation. As $\alpha_1(0) = \alpha_2(0) = \varphi_s(x)$, we obtain $\alpha_1 = \alpha_2$, proving the desired result.

It remains to show that the map $x \rightsquigarrow \varphi_t(x)$ is a diffeomorphism $\varphi_t : U \longrightarrow \varphi_t(U)$. Certainly φ_o is the identity transformation. We know that $\varphi_t(x)$ depends differentiably on x. For sufficiently small t and $x \in U$ therefore $\varphi_t(\varphi_{-t}(x)) = \varphi_{-t}(\varphi_t(x)) = \varphi_o(x) = X$, proving that φ_t is a diffeomorphism.

DEFINITION 5.1.6. A vectorfield A on X is <u>complete</u>, if it is induced by a 1-parameter group of transformations.

Example 5.1.7. Consider the vectorfield A on $\mathbb{R}$ which has in every point a positively oriented vector of length one. Consider the submanifold $(0, 1) \subset \mathbb{R}$. Then A restricted to $(0, 1)$ is not complete.

A criteria for completeness is given in

LEMMA 5.1.7. Let A be a vectorfield on X. Suppose there exists $\epsilon > 0$ and a local 1-parameter group of local transformations $\varphi: I_\epsilon \times X \longrightarrow X$ inducing A. Then φ has an extension to a 1-parameter group of transformations and A is therefore complete.

Proof: φ_t is a diffeomorphism for $|t| \lneqq \epsilon$. There is only to define φ_t for $|t| > \epsilon$. Write $t = k \cdot \frac{\epsilon}{2} + r$ with an integer k and $|r| < \frac{\epsilon}{2}$. If $k > 0$, define $\varphi_t = (\varphi_{\frac{\epsilon}{2}})^k \circ \varphi_r$. If $k < 0$, define $\varphi_t = (\varphi_{-\frac{\epsilon}{2}})^{-k} \circ \varphi_r$. Now φ_t satisfies all conditions.

Example 5.1.8. Let X be a compact manifold. Any vectorfield A on X is complete.

We remark that the relation between vectorfields and local 1-parameter groups of local transformations described in this section is at the origin of the denomination of a vectorfield as an infinitesimal transformation.

5.2. 1-parameter groups of transformations and equivariant maps.

Convention on notations. Given a local 1-parameter group of local transformations on X we denote it Ψ_t and just speak of the induced vectorfield A on X, without specifying the domains of definition. Also, for a given vectorfield A on X, we just write Ψ_t for a local 1-parameter group of local transformations inducing A on some subset of X. The formulas are valid as soon as they make sense. They make sense in particular if only 1-parameter group of transformations occur.

PROPOSITION 5.2.1. *Let Ψ_t, Ψ'_t be local 1-parameter groups of transformations on X, X', A, A' the induced vectorfields and $\varphi: X \to X'$ a map. If $\varphi \circ \Psi_t = \Psi'_t \circ \varphi$ for all t, then A and A' are φ-related.*

Proof: We have

$$\varphi_{*\Psi_t(x)} A_{\Psi_t(x)} = A'_{\Psi'_t(\varphi(x))}$$

by differentiating with respect to t. Or

$$\varphi_{*\Psi_t(x)} A_{\Psi_t(x)} = A'_{\varphi(\Psi_t(x))} ,$$

which shows the proposition in view of lemma 4.4.3.

It is convenient to call a map $\varphi: X \to X'$ satisfying $\varphi \circ \Psi_t = \Psi'_t \circ \varphi$ an equivariance with respect to the given local 1-parameter groups of local transformations Ψ_t and Ψ'_t. The proposition says that the induced vectorfields A and A' are then φ-related. This is characteristic for equivariant maps. Precisely we have

PROPOSITION 5.2.2. *Let X, X' be manifolds, A, A' vectorfields on X, X' respectively, Ψ_t, Ψ'_t corresponding local 1-parameter groups of local transformations and $\varphi: X \to X'$ a map. If A and A' are φ-related, then*

$$\varphi \circ \Psi_t = \Psi'_t \circ \varphi .$$

Proof: For $x \in X$ write $\alpha_1(t) = \varphi(\Psi_t(x))$, $\alpha_2(t) = \Psi'_t(\varphi(x))$. Then $\alpha_1(0) = \alpha_2(0) = \varphi(x)$. We prove $\alpha_1 = \alpha_2$ by showing that α_1, α_2 satisfy the same differential equation. But

$$\dot{\alpha}_1(t) = \varphi_{*_{\Psi_t(x)}} A_{\Psi_t(x)} = A'_{\alpha_1}(t)$$

by lemma 4.4.3, as A and A' are φ-related and

$$\dot{\alpha}_2(t) = A'_{\alpha_2}(t) \quad , \text{q.e.d.}$$

COROLLARY 5.2.3. Let X, X' be manifolds and $\varphi : X \longrightarrow X'$ a diffeomorphism. If A is a vectorfield on X, generating a local 1-parameter group of local transformations Ψ_t, then the vectorfield $\varphi_* A$ on X' generates the local 1-parameter group of local transformation $\varphi \circ \Psi_t \circ \varphi^{-1}$.

Proof: It is sufficient to observe that A and $\varphi_* A$ are φ-related.

Applying this to an automorphism we obtain

COROLLARY 5.2.4. Let X be a manifold, A a vectorfield on X generating a local 1-parameter group of local transformations Ψ_t and $\varphi : X \longrightarrow X$ a diffeomorphism. Then $\varphi_* A = A$ if and only if $\varphi \circ \Psi_t = \Psi_t \circ \varphi$ for all t.

We observe that the preceding is still true for a local automorphism of X, i.e. a map $\varphi : U \longrightarrow X$ defined on an open subset of X and having a restriction being a diffeomorphism. $\varphi_* A$ is then to be interpreted as a vectorfield on a convenient subset of X and can be defined by the formula of lemma 4.2.8. Consider in particular Ψ_s, which is a local automorphism of X in this sense. As $\Psi_s \circ \Psi_t = \Psi_t \circ \Psi_s$ for all t, we see that $(\Psi_s)_* A = A$ by corollary 5.2.4. This just means that the velocity field A of the flow Ψ_t is invariant by the flow, the characteristic property of a stationary flow.

An application of corollary 5.2.4 is the following:

LEMMA 5.2.5. Let G be a Lie group, $A \in LG$ and Ψ_t a local 1-parameter group of local transformations generated by A. Then

$$L_g \circ \Psi_t = \Psi_t \circ L_g \quad \text{for all } t,\ g \in G .$$

We can now prove

PROPOSITION 5.2.6. Let G be a Lie group. Every left invariant vectorfield on G is complete.

Proof: We consider a local 1-parameter group of local transformations Ψ_t generated by A, and show that Ψ_t has an extension to a local 1-parameter group of local transformations $\tilde{\Psi}: I_\epsilon \times G \longrightarrow G$ for an $\epsilon > 0$. Then the proposition follows by lemma 5.1.7. Suppose $\Psi: I_\epsilon \times U \longrightarrow G$ for an $\epsilon > 0$ and a neighborhood U of $e \in G$. A necessary condition for an extension $\tilde{\Psi}: I_\epsilon \times G \longrightarrow G$ is

$$\tilde{\Psi}_t(g) = (\tilde{\Psi}_t \circ L_g)(e) = (L_g \circ \tilde{\Psi}_t)(e) = L_g(\Psi_t(e))$$

by lemma 5.2.5. Defining conversely $\tilde{\Psi}$ by this formula, we obtain the desired $\tilde{\Psi}: I_\epsilon \times G \longrightarrow G$.

Another consequence of lemma 5.2.5 is

PROPOSITION 5.2.7. Let G be a Lie group, $A \in LG$ and Ψ_t the 1-parameter group of transformations generated by A. Then $\Psi_t = R_{\Psi_t(e)}$.

Proof: By lemma 5.2.5 a particular case of

LEMMA 5.2.8. Let G be a group (in the sense of algebra). A map $\Psi: G \longrightarrow G$ is a right translation (and then necessarily $R_{\Psi(e)} = \Psi$) if and only if

$$L_g \circ \Psi = \Psi \circ L_g \qquad \underline{\text{for all}}\ g \in G .$$

Proof: The associativity shows that the condition is necessary. Suppose conversely $L_g \circ \Psi = \Psi \circ L_g$ for all $g \in G$. For $\gamma \in G$ $g\Psi(\gamma) = \Psi(g\gamma)$ and in particular $g\Psi(e) = \Psi(g)$, i.e.

$$\Psi(g) = R_{\Psi(e)}\, g \quad , \text{ q.e.d.}$$

5.3. The bracket of two vectorfields.

We give now an interpretation of the bracket of two vectorfields (taken from K. Nomizu and S. Kobayashi [11], p. 15).

We use the notation $\varphi_* A$ for a vectorfield A on X and a local automorphism φ of X as explained in 5.2.

PROPOSITION 5.3.1. Let A and B be vectorfields on the manifold X, and φ_t a local 1-parameter group of local transformations generated by A. Then

$$[A, B]_x = \lim_{t \to 0} \frac{1}{t} [B_x - ((\varphi_t)_* B)_x] \qquad \text{for } x \in X$$

LEMMA 5.3.2. Let $\epsilon > 0$, $I_\epsilon = (-\epsilon, \epsilon) \subset \mathbb{R}$ and $f : I_\epsilon \times X \longrightarrow \mathbb{R}$ with $f(o, x) = 0$ for $x \in X$. Then there exists $g : I_\epsilon \times X \longrightarrow \mathbb{R}$ with $f(t, x) = tg(t, x)$. Moreover $g(o, x) = \frac{\partial f}{\partial t}(o, x)$.

Proof: Define $g(t, x) = \int_0^1 \frac{\partial f}{\partial t}(ts, x)\, ds$.

LEMMA 5. 3. 3. *Let* A *generate* φ_t. *For any* $f \in CX$ *there exists* $g_t \in CX$ *with* $f \circ \varphi_t = f + t g_t$ *and* $g_o = Af$.

The function $g(t, x) = g_t(x)$ *is defined, for each fixed* $x \in X$, *in* $|t| < \epsilon$ *for some* $\epsilon > 0$.

Proof: Consider $h(t, x) = f(\varphi_t(x)) - f(x)$ and apply lemma 5. 3. 2. Then $f \circ \varphi_t = f + tg_t$. We have

$$(Af)(x) = \lim_{t \to 0} \frac{1}{t}[f(\varphi_t(x)) - f(x)] = \lim_{t \to 0} \frac{1}{t} f(t, x)$$

$$= \lim_{t \to 0} g(t, x) = g_o(x) \quad .$$

Proof of proposition 5. 3. 1. Let $f \in CX$. Take $g_t \in CX$ as in lemma 5. 3. 3. Set $x_t = \varphi_t^{-1}(x)$. Then

$$((\varphi_t)_* B)_x f = (B(f \circ \varphi_t))(x_t) = (Bf)(x_t) + t(Bg_t)(x_t)$$

and

$$\lim_{t \to 0} \frac{1}{t}[B_x - ((\varphi_t)_* B)_x] f = \lim_{t \to 0} \frac{1}{t}[(Bf)(x) - (Bf)(x_t)] - \lim_{t \to 0} (Bg_t)(x_t)$$

$$= A_x(Bf) - B_x g_o = A_x(Bf) - B_x(Af)$$

$$= [A, B]_x f \quad , \quad \text{q. e. d.}$$

COROLLARY 5. 3. 4. *Let* A *and* B *be vectorfields on the manifold* X *and* φ_t *a local 1-parameter group of local transformations generated by* A. *Then for any value of* $s \in \mathbb{R}$, $x \in X$

$$((\varphi_s)_*[A,B])_x = \lim_{t\to 0} \frac{1}{t}[((\varphi_s)_* B)_x - ((\varphi_{t+s})_* B)_x]$$

Proof: Let $s \in \mathbb{R}$. Then $(\varphi_s)_*[A,B] = [(\varphi_s)_* A, (\varphi_s)_* B] = [A, (\varphi_s)_* B]$ as $(\varphi_s)_* A = A$ by the remark at the end of 5.2.

Applying proposition 5.3.1., we obtain

$$[A,(\varphi_s)_* B]_x = \lim_{t\to 0} \frac{1}{t}[((\varphi_s)_* B)_x - ((\varphi_t)_*(\varphi_s)_* B)_x]$$

$$= \lim_{t\to 0} \frac{1}{t}[((\varphi_s)_* B)_x - ((\varphi_{t+s})_* B)_x] .$$

PROPOSITION 5.3.5. Let X be a manifold, A and B vectorfields on X generating local 1-parameter groups of local transformations φ_t and Ψ_t respectively. Then $\varphi_t \circ \Psi_s = \Psi_s \circ \varphi_t$ for every s and t if and only if $[A,B] = O$.

Proof: If $\varphi_t \circ \Psi_s = \Psi_s \circ \varphi_t$ for every s and t, $(\varphi_t)_* B = B$ by corollary 5.2.4. By proposition 5.3.1, $[A,B] = O$. Suppose conversely $[A,B] = O$. By corollary 5.3.4, $\frac{d}{dt}((\varphi_t)_* B)_x = O$ for any t. Therefore $(\varphi_t)_* B = B$ for every t and by corollary 5.2.4 φ_t commutes with every Ψ_s.

PROPOSITION 5.3.6. Let the vectorfields A and B of X generate local 1-parameter groups of local transformations φ_t and Ψ_t respectively. Suppose $[A,B] = O$. Then $\chi_t = \varphi_t \circ \Psi_t = \Psi_t \circ \varphi_t$ is a local 1-parameter group of local transformations and is generated by $A + B$.

Proof: Proposition 5.3.5 shows that χ_t is indeed a local 1-parameter group of local transformations. Now

$$\dot{\chi}_t(x) = \dot{\varphi}_t(\Psi_t(x)) + (\varphi_t)_{*\Psi_t(x)} \dot{\Psi}_t(x)$$

$$= A_{\chi_t(x)} + (\varphi_t)_{*\Psi_t(x)} B_{\Psi_t(x)}$$

But by proposition 5.3.5 and corollary 5.2.4 $(\varphi_t)_* B = B$. Therefore

$$(\varphi_t)_{*\Psi_t(x)} B_{\Psi_t(x)} = B_{\varphi_t(\Psi_t(x))} = B_{\chi_t(x)}$$

and

$$\dot{\chi}_t(x) = (A + B)_{\chi_t(x)} \quad , \quad \text{q.e.d.}$$

5.4. 1-parameter subgroups of a Lie group.

DEFINITION 5.4.1. A 1-parameter subgroup α of a Lie group G is a homomorphism $\alpha : \mathbb{R} \longrightarrow G$ of Lie groups.

Remark. Let X be a manifold and φ_t a 1-parameter group of transformations of X. One would like to consider $t \rightsquigarrow \varphi_t$ as a 1-parameter subgroup $\mathbb{R} \longrightarrow \mathrm{Aut}\, X$ of $\mathrm{Aut}\, X$. But it does not make sense to speak of the differentiability of this map. See also the remark after example 3.2.5.

The trivial homomorphism $O : \mathbb{R} \longrightarrow G$ is a 1-parameter subgroup of G.

A non-trivial 1-parameter subgroup $\alpha: \mathbb{R} \longrightarrow G$ is not necessarily an injection, as shown by

Example 5.4.2. The canonical homomorphism $\mathbb{R} \longrightarrow \mathbb{R}/\mathbb{Z} = \mathbb{T}$ is a 1-parameter subgroup of $\mathbb{T}$.

Lemma 5.4.5 below shows that a non-trivial 1-parameter subgroup is an immersion.

Let A be a complete vectorfield on G and φ_t the 1-parameter group of transformations generated by A. Define

$$\alpha_t = \varphi_t(e) \qquad \text{for } e \in G$$

Then $\alpha: \mathbb{R} \longrightarrow G$ and $\alpha_o = \varphi_o(e) = e$, but α is not necessarily a 1-parameter subgroup of G. If $A \in LG$, this is the case, as stated in

PROPOSITION 5.4.3. Let G be a Lie group, $A \in LG$, φ_t the 1-parameter subgroup of G generated by A and $\alpha: \mathbb{R} \longrightarrow G$ the map defined by $\alpha_t = \varphi_t(e)$. Then α is a 1-parameter subgroup of G. Moreover $\varphi_t = R_{\alpha_t}$ and φ_t is completely described by α.

Proof: Applying lemma 5.2.5, we obtain

$$\alpha_{t_1+t_2} = \varphi_{t_1+t_2}(e) = \varphi_{t_1}(\varphi_{t_2}(e)) = (\varphi_{t_1} \circ L_{\varphi_{t_2}(e)})(e)$$

$$= (L_{\varphi_{t_2}(e)} \circ \varphi_{t_1})(e) = \varphi_{t_2}(e)\varphi_{t_1}(e) = \alpha_{t_2}\alpha_{t_1} .$$

In view of proposition 5.2.7 we have $\varphi_t = R_{\alpha_t}$. ▮

The statement $\varphi_t = R_{\alpha_t}$ is often paraphrased in the literature by:

"the infinitesimal transformation generated by a left invariant vectorfield is a right translation".

We call α the 1-parameter subgroup of G defined by $A \in LG$. Let $\mathcal{L}G$ denote the set of 1-parameter subgroups of G. We have defined a map $\Phi : LG \longrightarrow \mathcal{L}G$.

LEMMA 5.4.4. Let $A \in LG$, $\alpha = \Phi(A) \in \mathcal{L}G$. Then α is the solution of the differential equation

$$\dot{\alpha}_t = A_{\alpha_t}$$

with initial condition $\alpha_o = e$.

Proof: $\dot{\alpha}_t = \dot{\varphi}_t(e) = A_{\varphi_t(e)} = A_{\alpha_t}$ by proposition 5.1.3 and $\alpha_o = \varphi_o(e) = e$, q.e.d.

This lemma gives a direct description of the map $\Phi : LG \longrightarrow \mathcal{L}G$ and shows its injectivity. We shall see that Φ is bijective. First we prove

LEMMA 5.4.5. Let $\alpha \in \mathcal{L}G$. Then

$$\dot{\alpha}_t = (L_{\alpha_t})_{*_e} \dot{\alpha}_o = (R_{\alpha_t})_{*_e} \dot{\alpha}_o .$$

Proof: By differentiating $\alpha_{t+s} = \alpha_t \alpha_s = \alpha_s \alpha_t$ with respect to s, we obtain

$$\dot{\alpha}_{t+s} = (L_{\alpha_t})_{*_{\alpha_s}} \dot{\alpha}_s = (R_{\alpha_t})_{*_{\alpha_s}} \dot{\alpha}_s ,$$

and for $s = 0$ the desired result.

THEOREM 5.4.6. Let G be a Lie group, LG its Lie algebra and $\mathcal{L}G$ the set of 1-parameter subgroups of G. For any $A \in LG$ we define $\Phi(A) = a \in \mathcal{L}G$ as the solution of $\dot{a}_t = A_{a_t}$ with initial condition $a_0 = e$. Then $\Phi : LG \longrightarrow \mathcal{L}G$ is bijective.

Proof: Let $a \in \mathcal{L}G$. If $a = \Phi(A)$ for some $A \in LG$, then necessarily $A_e = \dot{a}_0$, which shows injectivity of Φ. Defining conversely $A \in LG$ as the vectorfield with $A_e = \dot{a}_0$, we have $A_{a_t} = (L_{a_t})_* \dot{a}_0 = \dot{a}_t$ by lemma 5.4.5, which shows surjectivity of Φ.

Lemma 5.4.5 shows that we obtain by the same definition a bijection $RG \longrightarrow \mathcal{L}G$. In fact, the tangent vectors of the curve $t \rightsquigarrow a_t$ belong as well to the left as to the right invariant vectorfield defining a. The situation is described precisely by

PROPOSITION 5.4.7. Let $A \in LG$, $a = \Phi(A) \in \mathcal{L}G$ and $g \in G$. Then $A_g = (ga_t)^{\bullet}_{t=0}$.

Proof: Let φ_t be the 1-parameter group of transformations generated by A. Then $\dot{\varphi}_t(g) = A_{\varphi_t(g)}$ for any $g \in G$. Now by proposition 5.4.3,

$$\varphi_t(g) = R_{a_t}(g) = ga_t$$

and

$$\dot{\varphi}_t(g) = (ga_t)^{\bullet}.$$

For $t = 0$ this shows $A_{\varphi_0(g)} = A_g = (ga_t)^{\bullet}_{t=0}$, q.e.d.

A consequence of theorem 5.4.6 is the

PROPOSITION 5.4.8. Let I be an open interval of $\mathbb{R}$ containing O and $\alpha : I \longrightarrow G$ a local homomorphism of Lie groups. Then there exists a unique 1-parameter subgroup $\tilde{\alpha} : \mathbb{R} \longrightarrow G$ of G with $\tilde{\alpha}/I = \alpha$.

Proof: Let $A \in LG$ be defined by $\dot{\alpha}_o = A_e$ and $\tilde{\alpha} = \Phi(A)$. Lemma 5.4.5 still applies to α, showing $\dot{\alpha}_t = (L_{\alpha_t})_{*_e} \dot{\alpha}_o$ and therefore $A_{\alpha_t} = (L_{\alpha_t})_{*_e} \dot{\alpha}_o = \dot{\alpha}_t$. But $\tilde{\alpha}$ is also a solution of this differential equation. Now $\alpha_o = \tilde{\alpha}_o = e$ shows $\tilde{\alpha}/I = \alpha$, and $\tilde{\alpha}$ is an extension of α to a 1-parameter subgroup of G. The uniqueness follows from the fact that there exists only one 1-parameter subgroup α of G with given $\dot{\alpha}_o$.

As shown by the example of the local isomorphism $\mathbb{T} \longrightarrow \mathbb{R}$, there exists not necessarily an extension of a local homomorphism $G \longrightarrow G'$ of Lie groups to a homomorphism (see remark following lemma 4.5.8). It follows from the theory of topological groups, that an extension exists, if G is simply connected (see also lemma 7.2.5). Proposition 5.4.8 is a particularly simple case of this situation.

LEMMA 5.4.9. Let G be a Lie group, A and $B \in LG$, α and β the corresponding elements $\in \mathcal{L}G$, φ_t and Ψ_t the 1-parameter groups of transformations generated by A and B respectively. Then $\varphi_t \circ \Psi_s = \Psi_s \circ \varphi_t$ for every t and s if and only if $\alpha_t \beta_s = \beta_s \alpha_t$ for every t and s.

Proof: For $g \in G$,

$$(\varphi_t \circ \Psi_s)(g) = (\varphi_t \circ \Psi_s \circ L_g)(e) = (L_g \circ \varphi_t \circ \Psi_s)(e) \text{ by lemma 5.2.5.}$$

Now $\Psi_s(e) = \beta_s = L_{\beta_s}(e)$ and therefore for the same reason

$$(L_g \circ \varphi_t \circ \Psi_s)(e) = (L_g \circ \varphi_t \circ L_{\beta s})(e) = (L_g \circ L_{\beta_s} \circ \varphi_t)(e)$$

thus proving

$$(\varphi_t \circ \Psi_s)(g) = g\beta_s \alpha_t \quad .$$

Similarly

$$(\Psi_s \circ \varphi_t)(g) = g\alpha_t \beta_s \quad .$$

This proves the lemma.

Consider the expression $(\varphi_t \circ \Psi_s)(g) = g\beta_s \alpha_t$, which occurred in the proof. In particular $(\varphi_t \circ \Psi_t)(e) = \beta_t \alpha_t$. If $[A, B] = O$, by proposition 5. 3. 6 $\chi_t = \varphi_t \circ \Psi_t = \Psi_t \circ \varphi_t$ is a 1-parameter group of transformations, $\chi_t(e) = \beta_t \alpha_t = \alpha_t \beta_t$ a 1-parameter subgroup, and $A + B$ the corresponding vectorfield. Therefore

PROPOSITION 5. 4. 10. Let $A, B \in LG$ and $\alpha, \beta \in \mathcal{L}G$ the corresponding 1-parameter subgroups. Suppose $[A, B] = O$. Then $\alpha_t \beta_t = \beta_t \alpha_t = \gamma_t$ defines a 1-parameter subgroup and $A + B$ is the corresponding left invariant vectorfield.

Together with proposition 5. 3. 5 we obtain from lemma 5. 4. 9

PROPOSITION 5. 4. 11. Let $A, B \in LG$, φ_t and Ψ_t the 1-parameter groups of transformations generated by A and B respectively and α, β the corresponding 1-parameter subgroups of G. Then the following statements are equivalent:

1) $[A, B] = 0$

2) $\varphi_t \circ \Psi_s = \Psi_s \circ \varphi_t$ *for every* s *and* t

3) $\alpha_t \beta_s = \beta_s \alpha_t$ *for every* s *and* t .

We shall see in chapter 6 that these conditions are even equivalent to $\alpha_t \beta_t = \beta_t \alpha_t$ for every t (see proposition 6. 5. 3.).

Now we consider a homomorphism $\rho : G \longrightarrow G'$ of Lie groups. A 1-parameter subgroup $\alpha \in \mathcal{L} G$ gives by composition with ρ an element $\rho \circ \alpha \in \mathcal{L} G'$. If ρ is a local homomorphism, $\rho \circ \alpha$ is a local homomorphism $\mathbb{R} \longrightarrow G'$, but can by proposition 5. 4. 8 be uniquely extended to a 1-parameter subgroup of G', which we also denote by $\rho \circ \alpha$. The map $\mathcal{L}(\rho) : \mathcal{L} G \longrightarrow \mathcal{L} G'$ so defined is compatible with the map $\bar{\Phi}$ of theorem 5. 4. 6. More precisely we have

PROPOSITION 5. 4. 12. *Let* G, G' *be Lie groups and* $\rho : G \longrightarrow G'$ *a local homomorphism. Then the following diagram*

$$\begin{array}{ccc} LG & \xrightarrow{L(\rho)} & LG' \\ {\scriptstyle \bar{\Phi}_G} \downarrow & & \downarrow {\scriptstyle \bar{\Phi}_{G'}} \\ \mathcal{L}(G) & \xrightarrow{\mathcal{L}(\rho)} & \mathcal{L}(G') \end{array}$$

commutes, where $\mathcal{L}(\rho)$ *is the composition with* ρ *and* $\bar{\Phi}_G$, $\bar{\Phi}_{G'}$ *the maps of theorem* 5. 4. 6.

Proof: For $\alpha \in \mathcal{L} G$ we have $(\rho \circ \alpha)_t /_{t=0} = \rho_{*_e} \dot{\alpha}_0$. This means that the diagram (without the dotted line)

$$\begin{array}{ccc} G_e & \xrightarrow{\rho_{*_e}} & G'_e \\ {\scriptstyle \mathcal{H}_G}\big\downarrow & & \big\downarrow{\scriptstyle \mathcal{H}_{G'}} \\ LG & \overset{L(\rho)}{- - - - \rightarrow} & LG' \\ {\scriptstyle \vec{\Phi}_G}\big\downarrow & & \big\downarrow{\scriptstyle \vec{\Phi}_{G'}} \\ \mathcal{L}G & \xrightarrow{\mathcal{L}(\rho)} & \mathcal{L}G' \end{array}$$

commutes. $L(\rho)$ is defined by filling in the dotted line in the upper half. $\mathcal{H}_G$ being surjective, (in fact bijective), this proves the proposition.

Define the forget-functor $V : \mathcal{L}\mathfrak{A} \longrightarrow \mathrm{Ens}$, assigning to any Lie algebra its underlying set and to any Lie algebra homomorphism the corresponding map of the underlying sets. Proposition 5.4.12 states then that $\vec{\Phi} : V \circ L \longrightarrow \mathcal{L}$ is a natural transformation, in fact a natural equivalence.

5.5. Killing vectorfields.

In this section, the relation between 1-parameter subgroups of a Lie group G and 1-parameter groups of transformations of a G-manifold X is studied.

Let X be a G-manifold with respect to a homomorphism $\tau : G \longrightarrow \mathrm{Aut}\, X$ and $\alpha : \mathbb{R} \longrightarrow G$ a 1-parameter subgroup of G. The composed homomorphism

$$\mathbb{R} \xrightarrow{\alpha} G \xrightarrow{\tau} \mathrm{Aut}\, X$$

defines a 1-parameter group of transformations φ_t of X. Indeed, the map

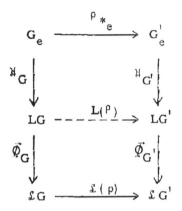

$$\mathbb{R} \times X \longrightarrow G \times X \longrightarrow X$$

$$(t, x) \rightsquigarrow (a_t, x) \rightsquigarrow \tau_{a_t}(x) = \varphi_t(x)$$

is differentiable.

DEFINITION 5.5.1. The Killing vectorfield A^* on X defined by $a \in \mathcal{L}G$ is the vectorfield induced by the 1-parameter group of transformations φ_t.

Remark. As already observed in section 5.4, one would like to consider the 1-parameter group of transformations φ_t as a 1-parameter subgroup $\mathbb{R} \longrightarrow \text{Aut } X$ of Aut X. Then φ_t would define a vectorfield on Aut X as explained in 5.4. As this presents difficulties, one proceeds as indicated above. The differential equation

$$\dot{\varphi}_t(x) = A^*_{\varphi_t(x)}$$

describing the relation between φ_t and A^* can heuristically be written as

$$\dot{\varphi}_t = A^*_{\varphi_t}$$

interpreting now A^* as a vectorfield on Aut X. This describes, of course, A^* only along the curve $t \rightsquigarrow \varphi_t$ on Aut X.

Example 5.5.2. If $\tau: \mathbb{R} \longrightarrow \text{Aut } X$ defines a 1-parameter group of transformations τ_t of X, then the Killing vectorfield A^* defined by $1_{\mathbb{R}}: \mathbb{R} \longrightarrow \mathbb{R}$ is just the vectorfield induced by τ_t.

PROPOSITION 5.5.3. Let G operate on G by left translations. The Killing vectorfield on G defined by $\alpha \in \mathcal{L}G$ is the right invariant vectorfield $B \in RG$ characterized by $\dot{\alpha}_o = B_e$.

Proof: The 1-parameter group of transformations $\varphi_t = L_{\alpha_t}$ on G induces by proposition 5.4.3 and theorem 5.4.6 (respectively the analogue for RG) a right invariant vectorfield B on G. It is characterized by $B_e = \dot{\alpha}_o$.

Example 5.5.4. Let $\rho : G \longrightarrow G'$ be a homomorphism and $\tau : G \longrightarrow \mathrm{Bij}\, G'$ the operation of G on G' defined by $\tau_g = L_{\rho(g)}$. Suppose $\alpha \in \mathcal{L}G$ and consider the 1-parameter group of transformations of G' defined by $\alpha : \varphi_t = L_{\rho(\alpha_t)} = L_{(\rho \circ \alpha)_t}$. It induces the right invariant vectorfield B' on G' characterized by

$$B'_{e'} = (\rho \circ \alpha)(t)^{\bullet}/_{t=0} = \rho_{*e}\, \dot{\alpha}_o .$$

Composing the correspondence $\alpha \rightsquigarrow B'$ with the canonical map $RG \longrightarrow \mathcal{L}G$, we obtain obviously just the homomorphism $R(\rho) : RG \longrightarrow RG'$ defined by $\rho : G \longrightarrow G'$.

Example 5.5.5. Consider a finite dimensional $\mathbb{R}$-vectorspace V and the natural representation of GL(V) in V. Let α be a 1-parameter subgroup of GL(V) and $v \in V$. Then the Killing vectorfield A^* on V defined by α satisfies $A^*_v = \dot{v}_o$, where $v_t = \alpha_t v$. But $\dot{v}_t = \dot{\alpha}_t v = \dot{\alpha}_o \alpha_t v$ by lemma 5.4.5. Therefore the Killing vectorfield A^* defined by the 1-parameter subgroup α satisfies $A^*_v = \dot{\alpha}_o v$, i.e. is the vectorfield canonically defined by the endomorphism $\dot{a}_o \in \mathcal{L}(V)$.

We now apply the results of section 5.3 to Killing vectorfields and prove

PROPOSITION 5.5.6. _Let_ X _be a_ G_-manifold with respect to a homomorphism_ $\tau: G \longrightarrow \operatorname{Aut} X$, α _a 1-parameter subgroup of_ G _and_ A^* _the Killing vectorfield on_ X _defined by_ α. _If_ C _is an arbitrary vectorfield on_ X, _then_

$$[A^*, C]_x = \lim_{t \to 0} \frac{1}{t}[C_x - ((\tau_{\alpha_t})_* C)_x] \qquad \text{for } x \in X .$$

Proof: $(\tau \circ \alpha)_t$ is the 1-parameter group of transformations of X generated by A^*. The formula is therefore a particular case of proposition 5.3.1.

COROLLARY 5.5.7. _Let_ G _be a Lie group_, $\alpha \in \mathcal{L}G$ _and_ $B \in RG$ _the corresponding right invariant vectorfield. If_ C _is an arbitrary vectorfield on_ G, _then_

$$[B, C]_g = \lim_{t \to 0} \frac{1}{t}[C_g - ((L_{\alpha_t})_* C)_g] \qquad \text{for } g \in G .$$

Proof: Let G operate on G by left translations. By proposition 5.5.3 B is then the Killing vectorfield defined by $\alpha \in \mathcal{L}G$ with respect to this operation. Now we are in the situation of proposition 5.5.6. ▮

Note that in particular for $C \in RG$ this formula expresses the bracket in RG with the aid of L_{α_t}. Of course, we have a similar formula for left invariant vectorfields. We deduce the following interesting formula:

PROPOSITION 5. 5. 8. Let G be a Lie group and $A, C \in LG$. If $\alpha \in \mathcal{L}G$ is the 1-parameter subgroup defined by A, then

$$[A, C] = \frac{d}{dt}\{Ad(\alpha_t)\}/_{t=0} C$$

where $Ad : G \longrightarrow \text{Aut } LG$ denotes the adjoint representation of G.

Proof: As in corollary 5. 5. 7, we first obtain for $g \in G$

$$[A, C]_g = \lim_{t \to 0} \frac{1}{t}[C_g - ((R_{\alpha_t})_* C)_g]$$

Now $C \in LG$ and therefore $(L_{\alpha_t}^{-1})_* C = C$, i. e.

$$(R_{\alpha_t})_* C = Ad(\alpha_t^{-1}) C .$$

This shows

$$[A, C]_g = -\frac{d}{dt}\{Ad(\alpha_t^{-1}) C_g\}/_{t=0}$$

which can be written

$$[-A, C]_g = \frac{d}{dt}\{Ad(\alpha_t^{-1}) C_g\}/_{t=0} .$$

But the subgroup $t \rightsquigarrow \alpha_t^{-1} = \alpha_{-t}$ corresponds to the vectorfield $-A \in LG$, showing thus the desired result. ▮

We have supposed $Ad : G \longrightarrow \text{Aut } LG$ to be a homomorphism of Lie groups. This follows from the continuity of Ad (see section 6. 3).

5.6. *The homomorphism* $\sigma: RG \longrightarrow DX$ *for a* G*-manifold* X.

The knowledge of the following two sections is not necessary for the understanding of the subsequent developments.

We shall show that an operation $\tau: G \longrightarrow \operatorname{Aut} X$ defines a homomorphism $\sigma: RG \longrightarrow DX$ of Lie algebras. First we prove

LEMMA 5.6.1. *Let* X *be a* G*-manifold,* X' *a* G'*-manifold,* $\rho: G \longrightarrow G'$ *a homomorphism and* $\varphi: X \longrightarrow X'$ *a* ρ*-equivariance. Consider* $\alpha \in \mathcal{L}G$, $\alpha' = \rho \circ \alpha \in \mathcal{L}G'$ *and the Killing vectorfields* A^*, A'^* *defined by* α, α'. *Then* A^* *and* A'^* *are* φ*-related.*

Proof: Let Ψ_t, Ψ'_t be the 1-parameter groups of transformations of X, X' defined by α, α' : $\Psi_t = \tau_{\alpha_t}$, $\Psi'_t = \tau'_{\alpha'_t}$. It is sufficient to prove

$$\varphi \circ \Psi_t = \Psi'_t \circ \varphi$$

in view of proposition 5.2.1.

Now the ρ-equivariance of φ signifies the commutativity of the diagram

$$\begin{array}{ccc} G \times X & \xrightarrow{\ \tau\ } & X \\ {\scriptstyle \rho \times \varphi}\downarrow & & \downarrow{\scriptstyle \varphi} \\ G' \times X' & \xrightarrow{\ \tau'\ } & X' \end{array}$$

As $\alpha' = \rho \circ \alpha$, the following diagram is also commutative:

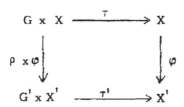

$$\begin{array}{ccc} \mathbb{R} \times X & \xrightarrow{\alpha \times 1_X} & G \times X \\ {\scriptstyle 1_{\mathbb{R}} \times \varphi} \downarrow & & \downarrow {\scriptstyle \rho \times \varphi} \\ \mathbb{R} \times X' & \xrightarrow{\alpha' \times 1_{X'}} & G' \times X' \end{array}$$

Composing these diagrams, we obtain the commutative diagram

$$\begin{array}{ccc} \mathbb{R} \times X & \xrightarrow{\Psi} & X \\ {\scriptstyle 1_{\mathbb{R}} \times \varphi} \downarrow & & \downarrow {\scriptstyle \varphi} \\ \mathbb{R} \times X' & \xrightarrow{\Psi'} & X' \end{array}$$

which proves $\varphi \circ \Psi_t = \Psi'_t \circ \varphi$. ∎

We are now in the position to prove the fundamental

THEOREM 5.6.2. *Let* G *be a Lie group,* X *a manifold,* RG *the Lie algebra of right invariant vectorfields on* G, *and* DX *the Lie algebra of vectorfields on* X. *An operation* $\tau : G \longrightarrow \mathrm{Aut}\, X$ *defining* X *as* G*-manifold induces a homomorphism*

$$\tau : RG \longrightarrow DX .$$

If $B \in RG$ *and* $\alpha \in \mathcal{L}\, G$ *the 1-parameter subgroup defined by* $\dot{\alpha}_t = B_{\alpha_t}$, *then* $\sigma(B)$ *is the Killing vectorfield on* X *defined by* α.

Proof: Let $B \in RG$, $\sigma(B) \in DX$. We show that B and $\sigma(B)$ are p-related under the effect of a map $p : G \longrightarrow X$. The theorem will then follow by lemma 4.4.2.

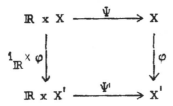

Choose $x_o \in X$ and define $p: G \longrightarrow X$ by $p(g) = \tau_g(x_o)$. Then the diagram

$$\begin{array}{ccc} G & \xrightarrow{L_g} & G \\ {\scriptstyle p}\downarrow & & \downarrow{\scriptstyle p} \\ X & \xrightarrow{\tau_g} & X \end{array}$$

is commutative. Considering the operation of G on G by left translations, this means that p is an equivariance. Let $a \in \mathcal{L}G$. The corresponding Killing vectorfields on G and X are p-related by lemma 5.6.1. The Killing vectorfield on G defined by a is by proposition 5.5.3 the element $B \in RG$. But the Killing vectorfield on X defined by a is just $\sigma(B)$ and therefore B and $\sigma(B)$ are p-related. ∎

Denote by KX the set of Killing vectorfields on the G-manifold X. Then $KX = \mathfrak{J}m\sigma$ and by theorem 5.6.2 KX is a Lie algebra.

Example 5.6.3. Let G operate on G by left translations. By proposition 5.5.3 the Lie algebra of Killing vectorfields is RG and the homomorphism $\sigma: RG \longrightarrow RG$ the identity 1_{RG}.

Example 5.6.4. Let $\rho: G \longrightarrow G'$ be a homomorphism and $\tau: G \longrightarrow \text{Bij } G'$ the operation of G on G' defined by $\tau_g = L_{\rho(g)}$. By example 5.5.4 the Lie algebra of Killing vectorfields is a subalgebra of RG' and the homomorphism $R(\rho): RG \longrightarrow RG'$ the homomorphism σ of theorem 5.6.2.

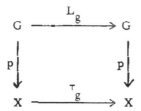

Example 5.6.5. Let V be a finite-dimensional $\mathbb{R}$-vectorspace and consider the natural representation of $G = GL(V)$ in V. We identify RG and LG with G_e by the canonical isomorphisms. We know from proposition 4.3.8 that then $LG = \mathcal{L}(V)$. But by corollary 4.6.8, we also have $RG = (LG)^o$. Therefore $RG = (\mathcal{L}(V))^o$ after identification with G_e. Now we see from example 5.5.5, that the homomorphism $\sigma: RG \longrightarrow KV$ in this case is just the canonical map $(\mathcal{L}(V))^o \longrightarrow DV$, assigning to any endomorphism $A \in \mathcal{L}(V)$ the vectorfield $A^*_v = Av$. That this map conserves the bracket, is seen directly as follows. Let $A_1, A_2 \in \mathcal{L}(V)$. Then their bracket in $(\mathcal{L}(V))^o$ is $A_2A_1 - A_2A_1$. On the other hand

$$[A_1^*, A_2^*]_v = (\frac{d}{dv}A^*_{2_v})(v)A^*_{1_v} - (\frac{d}{dv}A^*_{1_v})(v)A^*_{2_v}$$

by the same formula as in example 4.3.8.

But

$$(\frac{d}{dv}A^*_{2_v})(v)A^*_{1_v} = A^*_{2_v}A^*_{1_v} = A_2A_1v$$

and therefore

$$[A_1^*, A_2^*]_v = (A_2A_1 - A_1A_2)v \quad , \qquad \text{q.e.d.}$$

Consider more generally a representation of the Lie group G in V. The homomorphism $\tau: G \longrightarrow GL(V)$ induces a homomorphism $R(\tau): RG \longrightarrow R(GL(V)) = (\mathcal{L}(V))^o$ and the induced homomorphism $\sigma: RG \longrightarrow DV$ is just the composition of this homomorphism with the one $(\mathcal{L}(V))^o \longrightarrow DV$ described before.

We observe that the general situation for a G-manifold X is in some sense similar to that of example 5.6.5. The operation $\tau: G \longrightarrow \text{Aut } X$ induces by composition with the natural homomorphism (it's even an isomorphism) a representation $G \longrightarrow \text{Aut } CX$ of G in CX. The trouble is that CX is very big and one cannot just speak of the differentiability of this map. But if we don't care about this, we can view the induced homomorphism $RG \longrightarrow DX \longleftrightarrow OX$ into the Lie algebra of operators on X as the representation of RG in CX induced by the representation of G in CX, exactly as in example 5.6.5.

The Lie algebra of Killing vectorfields on a G-manifold is by definition a Lie algebra of complete vectorfields. It is natural to ask if any finite dimensional Lie algebra of complete vectorfields on a manifold X can be interpreted as the Lie algebra of Killing vectorfields corresponding to the operation of some Lie group G on X. We discuss the particular case of a commutative Lie algebra. See R. Palais [13] for a positive answer to our question in general.

Let a commutative Lie group G operate on X. As $\tau : RG \longrightarrow KX$ is a surjective homomorphism, the Lie algebra KX of Killing vectorfields is commutative.

We prove a converse

PROPOSITION 5.6.6. <u>Let</u> K <u>be a finite dimensional, commutative Lie algebra of complete vectorfields on</u> X. <u>Then there is an operation</u>

of the additive group of K on X, such that the Lie algebra K is the Lie algebra of Killing vectorfields of this operation.

Proof: Consider the additive structure on K, making it a Lie group with Lie algebra K. We define an operation of K on X as follows. Let $A \in K$ and φ_t the 1-parameter group of transformations of X generated by A. Then we set $\tau_A = \varphi_1$, defining thus a map $\tau: K \longrightarrow \mathrm{Aut}\, X$. We show τ to be a homomorphism.

Let $A, B \in K$ generate φ_t, Ψ_t respectively. Then by proposition 5.3.6, $\chi_t = \varphi_t \circ \Psi_t$ is the 1-parameter group of transformations generated by $A + B$. Therefore $\tau_{A+B} = \chi_1 = \varphi_1 \circ \Psi_1 = \tau_A \circ \tau_B$, as was to be proved.

We observe that $\tau_{tA} = \varphi_t$. Now the 1-parameter subgroup of K corresponding to A is $t \leadsto tA$ and therefore the corresponding Killing vectorfield A^* on X satisfies

$$A^*_x = \frac{d}{dt}\{\tau_{tA}(x)\}\Big/_{t=0} = \frac{d}{dt}\varphi_t(x)\Big/_{t=0} = A_x .$$

This shows that the homomorphism of theorem 5.6.2 is just the identity in this case. This finishes the proof.

The homomorphism $\sigma: RG \longrightarrow DX$ defined for a G-manifold X in theorem 5.6.2 reflects particular properties of the operation $\tau: G \longrightarrow \mathrm{Aut}\, X$. Namely

PROPOSITION 5.6.7. Let $\tau: G \longrightarrow \mathrm{Aut}\, X$ define X as a G-manifold and $\sigma: RG \longrightarrow DX$ be the induced homomorphism.

(i) <u>If τ is injective (i.e. an effective operation, then σ is injective.</u>

(ii) <u>If τ is a free operation, then a Killing vectorfield is either everywhere zero or nowhere zero.</u>

<u>Proof:</u> Let $B \in RG$ and $\alpha \in \mathcal{L}G$ the corresponding 1-parameter subgroup. Then $(\sigma B)_{\varphi_t(x)} = \dot{\varphi}_t(x)$, where $\varphi_t = \tau_{\alpha_t}$.

(i) Suppose $\sigma B = O$. Then $\dot{\varphi}_t(x) = O$ and $\varphi_t(x) = x$ for every x and t. τ being injective, $\tau_{\alpha_t} = 1_X$ implies $\alpha_t = e$ and $B_e = \dot{\alpha}_o = O$, i.e. $B = O$.

(ii) Suppose $(\sigma B)_x = O$ for some $x \in X$. Then $\dot{\varphi}_o(x) = O$ and $\varphi_t(x) = x$ for every t (x fixed) because of the uniqueness of the solution of $\dot{\varphi}_t(x) = (\sigma B)_{\varphi_t(x)}$ with $\varphi_o(x) = x$. τ being free, $\varphi_t(x) = \tau_{\alpha_t}(x) = x$ for some x implies $\alpha_t = e$ and as before $B = O$ and $\sigma B = O$. (Remember that a free operation is injective, so by (i) $B = O$ and $\sigma B = O$ are equivalent statements.)

Note that the injectivity of $\sigma : RG \longrightarrow DX$ does not imply the injectivity of $\tau : G \longrightarrow \mathrm{Aut}\, X$.

Example 5.6.8. A homomorphism $\varrho : G \longrightarrow G'$ induces an operation $\tau = L \circ \varrho$ of G on G' (see example 5.6.4) and the induced homomorphism σ is just $R(\varrho) : RG \longrightarrow RG'$. $R(\varrho)$ can be injective without $\tau = L \circ \varrho$ being injective. It is sufficient to exhibit a non-injective $\varrho : G \longrightarrow G'$ with injective $\varrho_{*_e} : G_e \longrightarrow G'_e$. The canonical homomorphism $\mathbb{R} \longrightarrow \mathbb{R}/\mathbb{Z} = \mathbb{T}$ is such a case.

Remark. Let us discuss the results of this section from the heuristic point of view already mentioned several times, considering Aut X as a Lie group. This is an effective operation on X and defines therefore by 5.6.7 an injection $R(\text{Aut } X) \xrightarrow{\tilde{\sigma}} DX$. It is natural to think that the image is the set of all complete vectorfields on X. But this set is not necessarily a Lie algebra, which destroys many hopes. Otherwise one would decrete this algebra to be R(Aut X). In case X is compact, there is however no problem, every vectorfield being complete. Now any operation $\tau: G \longrightarrow \text{Aut } X$ can be thought to define a homomorphism $R(\tau) : RG \longrightarrow R(\text{Aut } X)$. Then the homomorphism $\sigma: RG \longrightarrow DX$ of theorem 5.6.2 is just the composition $\sigma = \tilde{\sigma} \circ R(\tau)$.

Exercise 5.6.9. Let G be a commutative group operating on the manifold X and KX be the Lie algebra of Killing vectorfields on X. Show that every element of KX is invariant under the action of G on KX.

5.7. Killing vectorfields and equivariant maps.

We shall study the compatibility with equivariant maps of the homomorphism $\sigma : RG \longrightarrow DX$ defined in section 5.6 for a left operation. It is clear that considering a right operation of G on X, one obtains similarly a homomorphism $\sigma: LG \longrightarrow DX$.

We prove first the

LEMMA 5.7.1. Let X, X' be manifolds, $\varphi: X \longrightarrow X'$ a map and A, A_1', A, A_2' pairs of φ-related vectorfields on X, X'. If φ is *surjective, then* $A_1' = A_2'$.

Proof: φ^* is injective, because $\varphi^* f_1 = \varphi^* f_2$ or $f_1 \circ \varphi = f_2 \circ \varphi$ implies $f_1 = f_2$ if φ is surjective. But by definition of φ-relatedness, we have commutative diagrams

$$\begin{array}{ccc} CX & \xleftarrow{\varphi^*} & CX \\ {\scriptstyle A}\uparrow & & \uparrow{\scriptstyle A_i'} \\ CX & \xleftarrow{\varphi^*} & CX \end{array} \qquad (i = 1, 2)$$

Injectivity of φ^* clearly implies $A_1' = A_2'$.

PROPOSITION 5.7.2. *Let* X *be a* G^o-*manifold with respect to a right operation* $\tau: G^o \longrightarrow \mathrm{Aut}\, X$, X' *a* G'^o-*manifold with respect to a right operation* $\tau': G'^o \longrightarrow \mathrm{Aut}\, X'$, $\rho: G \longrightarrow G'$ *a homomorphism and* $\varphi : X \longrightarrow X'$ *a* ρ-*equivariant map. Consider the induced homomorphisms* $\sigma: LG \longrightarrow KX \hookrightarrow DX$, $\sigma' : LG' \longrightarrow KX' \hookrightarrow DX'$ *of theorem* 5.6.2. *If either* G *operates effectively on* X *or* φ *is surjective, there is a unique map* $\gamma: KX \longrightarrow KX'$ *making commutative the diagram*

$$\begin{array}{ccc} LG & \xrightarrow{\sigma} & KX \\ {\scriptstyle L(\rho)}\downarrow & & \downarrow{\scriptstyle \gamma} \\ LG' & \xrightarrow{\sigma'} & KX' \end{array}$$

and this map is a homomorphism of Lie algebras.

Proof: Let $A \in LG$. By proposition 5.6.1, the vectorfields $A^* = \sigma(A)$ and $\sigma'(L(\rho)A)$ are φ-related.

Suppose first that G operates effectively on X, i.e. that $\tau: G^o \longrightarrow \operatorname{Aut} X$ is injective. By proposition 5.6.7, σ is then also injective. For $A^* \in KX$ there is therefore a unique $A \in LG$ with $\sigma(A) = A^*$. We define $\gamma(A^*) = \sigma'(L(\rho)A)$.

Suppose now that φ is surjective and let $A \in LG$. As $A^* = \sigma(A)$ and $\sigma'(L(\rho)A)$ are φ-related, by lemma 5.7.1, $\sigma'(L(\rho)A)$ is uniquely defined by A^*. We can therefore define γ as before.

The uniqueness of γ follows from the surjectivity of $\sigma: LG \longrightarrow KX$ and γ is a homomorphism by lemma 4.4.2, q.e.d.

COROLLARY 5.7.3. *Let the situation be as in proposition 5.7.2. If $\varphi: X \longrightarrow X'$ is a ρ-equivariant diffeomorphism, then the following diagram commutes.*

$$\begin{array}{ccccc}
LG & \longrightarrow & KX & \hookrightarrow & DX \\
\downarrow{\scriptstyle L(\rho)} & & \downarrow{\scriptstyle \gamma} & & \downarrow{\scriptstyle \varphi_*} \\
LG' & \longrightarrow & KX' & \hookrightarrow & DX'
\end{array}$$

Proof: If $C \in DX$, then C and $\varphi_* C$ are φ-related. Therefore $\varphi_*/KX = \gamma$, q.e.d.

We want to apply this to the diffeomorphism $\tau_{g^{-1}}: X \longrightarrow X$ defined by a right-operation $\tau: G^o \longrightarrow \operatorname{Aut} X$. First we remark

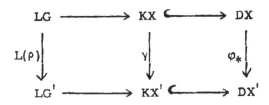

LEMMA 5.7.4. $\tau_{g^{-1}} : X \longrightarrow X$ is $\mathfrak{J}_g$-equivariant.

Proof: The diagram

$$\begin{array}{ccc} X & \xrightarrow{\tau_{g^{-1}}} & X \\ {\scriptstyle \tau_\gamma}\downarrow & & \downarrow{\scriptstyle \tau_{\mathfrak{J}_{g(\gamma)}}} \\ X & \xrightarrow{\tau_{g^{-1}}} & X \end{array}$$

commutes for $\gamma \in G : \tau_{\mathfrak{J}_{g(\gamma)}} \circ \tau_{g^{-1}} = \tau_{g\gamma g^{-1}} \circ \tau_{g^{-1}} = \tau_{\gamma g^{-1}} = \tau_{g^{-1}} \circ \tau_\gamma$.

By corollary 5.7.3 therefore we have

PROPOSITION 5.7.5. Let X be a G^o-manifold with respect to a right operation $\tau : G^o \longrightarrow \mathrm{Aut}\, X$ and $\sigma : LG \longrightarrow KX \longrightarrow DX$ the induced homomorphism of theorem 5.6.2. Then the following diagram commutes.

$$\begin{array}{ccccc} LG & \xrightarrow{\sigma} & KX & \hookrightarrow & DX \\ {\scriptstyle Adg}\downarrow & & & & \downarrow{\scriptstyle (\tau_{g^{-1}})_*} \\ LG & \xrightarrow{\sigma} & KX & \hookrightarrow & DX \end{array}$$

We just remember that $Ad\, g = L(\mathfrak{J}_g)$. Further it is clear that $(\tau_{g^{-1}})_* = \alpha_g$ defines DX as G-Lie algebra, τ being a right operation.
We obtain therefore

THEOREM 5.7.6. Let X be a G^o-manifold with respect to a homomorphism $\tau : G^o \longrightarrow \mathrm{Aut}\, X$. Consider the adjoint representation of G in LG and the operation of G on DX defined by $\alpha_g = (\tau_{g^{-1}})_*$. Then the induced homomorphism $\sigma : LG \longrightarrow DX$ is an equivariance.

For $A \in LG$ and the corresponding $A^* = \sigma(A)$ we have therefore the formula

$$(\tau_{g^{-1}})_* A^* = \sigma(\operatorname{Ad} g\, A) .$$

This shows in particular that τ_g transforms Killing vectorfields in Killing vectorfields.

Example 5.7.7. Let G operate on G by right translations. The commutativity of the diagram

$$\begin{array}{ccc} LG & \longrightarrow & DG \\ {\scriptstyle \operatorname{Ad} g}\downarrow & & \downarrow{\scriptstyle (R_{g^{-1}})_*} \\ LG & \longrightarrow & DG \end{array}$$

expressed by the theorem is just proposition 4.7.3.

For effective operations, σ is an injection by proposition 5.6.7. The commutative diagram

$$\begin{array}{ccc} LG & \xrightarrow{\sigma} & DX \\ {\scriptstyle \operatorname{Ad} g}\downarrow & & \downarrow{\scriptstyle a_g} \\ LG & \xrightarrow{\sigma} & DX \end{array}$$

shows that $a: G \longrightarrow \operatorname{Aut} DX$ can be interpreted as an extension of the adjoint representation of G.

LEMMA 5.7.8. *The homomorphism* $\tau: G^o \longrightarrow \operatorname{Aut} X$ *is injective if and only if the homomorphism* $a: G \longrightarrow \operatorname{Aut} DX$ *defined by*

$\alpha_g = (\tau_{g^{-1}})_*$ is injective.

Proof: α is the composition

$$\begin{array}{ccccccc} G & \xrightarrow{I} & G & \xrightarrow{\tau} & \text{Aut } X & \xrightarrow{*} & \text{Aut } DX \\ g & \rightsquigarrow & g^{-1} & \rightsquigarrow & \tau_{g^{-1}} & \rightsquigarrow & (\tau_{g^{-1}})_* \end{array}$$

I is bijective.

$*$ is injective, because $\varphi \in \text{Aut } X$ with $\varphi_* = 1_{DX}$ implies $\varphi_{*_x} = 1_{T_x(X)}$ and $\varphi = 1_X$. Therefore α is injective if and only if $* \circ \tau$ is injective if and only if τ is injective, q.e.d.

Remark. For a left-operation $\tau : G \longrightarrow \text{Aut } X$ we have similarly a commutative diagram

$$\begin{array}{ccc} RG & \xrightarrow{\sigma} & DX \\ {\scriptstyle R(\mathfrak{I}_g)} \downarrow & & \downarrow {\scriptstyle \delta_g} \\ RG & \xrightarrow{\sigma} & DX \end{array}$$

where $\delta_g = (\tau_g)_*$ and σ is therefore an equivariance with respect to the (left) operations of G on RG and DX. Let us take up again our heuristic viewpoint of looking at Aut X as a Lie group. Aut X operates naturally from the left on X, defining a homomorphism $\tilde{\sigma} : R(\text{Aut } X) \longrightarrow DX$ (see remark at the end of section 5.6). As just observed, for $\varphi \in \text{Aut } X$ we have a commutative diagram

$$\begin{array}{ccc} R(\mathrm{Aut}\ X) & \xrightarrow{\ \tilde{\sigma}\ } & DX \\ {\scriptstyle R(\mathcal{I}_\varphi)}\Big\downarrow & & \Big\downarrow{\scriptstyle \varphi_*} \\ R(\mathrm{Aut}\ X) & \xrightarrow{\ \tilde{\sigma}\ } & DX \end{array}$$

$\tilde{\sigma}$ being an injection, we see that $\varphi_* : DX \longrightarrow DX$ is an extension of $R(\mathcal{I}_\varphi)$, which is the adjoint representation of Aut X in R(Aut X). By the argument in the proof of lemma 5. 7. 8 we see that $\mathrm{Aut}\ X \xrightarrow{\ *\ } \mathrm{Aut}\ DX$, sending φ to φ_*, is injective. This is in accordance with lemma 5. 7. 8, $\tilde{\sigma}$ being an injection.

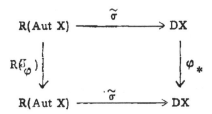

Chapter 6. THE EXPONENTIAL MAP OF A LIE GROUP

The relation between 1-parameter subgroups of a Lie group G and invariant vectorfields on G studied in section 5.4 is used to define a map $\exp: G_e \to G$, which turns out to have wonderful properties.

6.1. Definition and naturality of exp.

In the following, for convenience, we identify LG with G_e and write $A \in G_e$ for a tangent vector at e.

DEFINITION 6.1.1. The exponential map $\exp: G_e \to G$ is the map defined by

$$\exp A = \alpha_1 \qquad \text{for } A \in G_e$$

where α is the 1-parameter subgroup of G defined by A according to theorem 5.4.6. With our convention $A = \dot{\alpha}_0$.

Let $\alpha \in \mathcal{L}G$ and define for a $t \in \mathbb{R}$ $\beta_s = \alpha_{st}$. Then clearly $\beta \in \mathcal{L}G$ and moreover we have

LEMMA 6.1.2. $\dot{\beta}_0 = t\dot{\alpha}_0$.

Proof: For $f \in CG$ we have

$$\dot{\beta}_0 f = \frac{d}{ds} f(\beta_s)\Big|_{s=0} = \frac{d}{ds} f(\alpha_{st})\Big|_{s=0} = t\dot{\alpha}_0 f, \qquad \text{q.e.d.}$$

This shows that $\exp(tA) = \alpha_t$.

PROPOSITION 6.1.3. $\exp((t_1+t_2)A) = \exp(t_1A)\cdot\exp(t_2A)$

Proof: α is a homomorphism. ∎

The proposition 5.4.10 shows that for $A, B \in G_e$ with $[A, B] = 0$ we have the formula

$$\exp(t(A+B)) = \exp(tA) \cdot \exp(tB)$$

and in particular for $t = 1$

$$\exp(A+B) = \exp A \cdot \exp B \text{ , showing}$$

PROPOSITION 6.1.4. <u>If the Lie algebra</u> LG <u>of</u> G <u>is commutative, then</u> $\exp: G_e \to G$ <u>is a homomorphism of the additive vectorgroup</u> G_e <u>into</u> G.

To justify the notation exp, let us consider the

Example 6.1.5. Let V be a finite dimensional $\mathbb{R}$-vectorspace and $G = GL(V)$. In proposition 4.3.8 we have seen that identifying LG with G_e, we have $LG = \mathcal{L}(V)$. Now let $A \in \mathcal{L}(V)$ and $\alpha \in \mathcal{L}G$ the corresponding 1-parameter subgroup. We show that in this case

$$\exp(tA) = \alpha_t = e^{tA} = \sum_{n=0}^{\infty} \frac{1}{n!} (tA)^n .$$

To prove this, consider $\beta_t = e^{tA} = \Sigma_{n=0}^{\infty} \frac{1}{n!} (tA)^n$. Then $\dot{\beta}_t = \Sigma_{n=0}^{\infty} \frac{n}{n!} (tA)^{n-1} A = \beta_t A$ and $\beta_0 = 1_V$. But also $\dot{\alpha}_t = \alpha_t \dot{\alpha}_0 = \alpha_t A$ and $\alpha_0 = 1_V$. Therefore $\alpha = \beta$, as both satisfy the same differential equation with the same initial condition, q.e.d.

Consider in particular $V = \mathbb{R}$. Then $GL(V) = \mathbb{R}^*$, the multiplicative group of real numbers different from zero. The Lie algebra of $\mathbb{R}^*$ is $\mathbb{R}$ with the (only possible) trivial Lie algebra structure. The map $\exp: \mathbb{R} \to \mathbb{R}^*$ is just the ordinary exponential map.

We now show the naturality of exp, i.e.

PROPOSITION 6.1.6. Let $\rho: G \to G'$ be a local homomorphism. Then the following diagram (taken in the sense of local maps) is commutative.

$$\begin{array}{ccc} G_e & \xrightarrow{\rho_{*e}} & G'_e \\ \downarrow{\scriptstyle \exp} & & \downarrow{\scriptstyle \exp} \\ G & \xrightarrow{\rho} & G' \end{array}$$

Proof: Let $A \in G_e$. If $\alpha_t = \exp(tA)$ is the corresponding 1-parameter subgroup of G, then $(\rho \circ \alpha)(t)^{\cdot}\big|_{t=0} = \rho_{*e}\dot{\alpha}_0 = \rho_{*e} A$. Therefore

$$\exp(t\rho_{*e} A) = \rho(\exp(tA))$$

and for $t = 1$

$$\exp(\rho_{*e} A) = \rho(\exp A), \qquad \text{q.e.d.}$$

As an application, we obtain

COROLLARY 6.1.7. For $g \in G$, $A \in LG$ we have

$$\exp(\mathrm{Ad}g\, A) = g \exp A g^{-1}.$$

Proof: $\mathrm{Ad}g = L(\mathfrak{J}_g)$ by definition. Note that after identification of LG with G_e the adjoint representation operates in G_e.

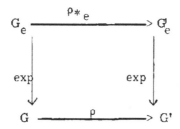

Another application of proposition 6.1.6 is the following. Consider a finite-dimensional $\mathbb{R}$-vectorspace V. The Lie algebra of $GL(V)$ is by proposition 4.3.8 equal to $\mathfrak{L}(V)$. Now the homomorphism $\det: GL(V) \to \mathbb{R}^*$ induces by proposition 4.5.11 the Lie algebra homomorphism $\operatorname{tr}: \mathfrak{L}(V) \to \mathbb{R}$. The naturality of the exponential mapping proves

COROLLARY 6.1.8. For any $A \in \mathfrak{L}(V)$

$$\det \exp A = \exp \operatorname{tr} A$$

The image of the map $\exp: G_e \to G$ is contained in G_0, the connected component of the identity of G. The following example shows that $\exp$ need not be surjective even for connected G.

Example 6.1.9. Let $SL(2, \mathbb{R})$ be the group of 2-rowed quadratic matrices with determinant 1. It is a connected Lie group. We show that there is an element in $SL(2, \mathbb{R})$ which is not a square. This will imply that $\exp$ is not surjective.

Let $g \in SL(2, \mathbb{R})$ and consider its characteristic polynomial $\det(\lambda \mathfrak{Id} - g) = \lambda^2 - \lambda \operatorname{tr} g + \det g$ where $\operatorname{tr} g$ denotes the trace of g. Now $\det g = 1$ and therefore, by a theorem of linear algebra, $g^2 - \operatorname{tr} g \cdot g + \mathfrak{Id} = 0$. Applying the trace function, we obtain $\operatorname{tr} g^2 = (\operatorname{tr} g)^2 - 2 \geqq -2$.

Consider the element

$$\ell = \begin{pmatrix} -2 & 0 \\ 0 & -1/2 \end{pmatrix} \quad \text{of } SL(2, \mathbb{R}).$$

As $\operatorname{tr} \ell < -2$, the equation $\ell = g^2$ has no solution.

Remark. Consider a manifold X and a 1-parameter group of transformations φ_t of X generated by a vectorfield A^* on X. Considering φ_t as a 1-parameter subgroup of $\operatorname{Aut} X$, it is suggestive to write as for Lie group

$$\varphi_t = \exp tA^*$$

View now, as before, an operation of G on X as a homomorphism

$$G \xrightarrow{\ \tau\ } \operatorname{Aut} X$$

and let $a \in \mathcal{L}G$ and A^* be the Killingvectorfield defined by a.
Then by definition

$$\tau_{a_t} = \exp tA^*$$

or if $A \in G_e$ with $a_t = \exp tA$

$$\tau_{\exp tA} = \exp tA^* \qquad .$$

This expresses just the commutativity of the diagram

$$\begin{array}{ccc} G_e & \xrightarrow{\ \sigma\ } & KX \\ {\scriptstyle \exp}\downarrow & & \downarrow{\scriptstyle \exp} \\ G & \xrightarrow{\ \tau\ } & \operatorname{Aut} X \end{array}$$

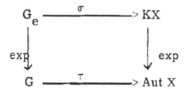

where $\sigma: G_e \to KX \hookrightarrow DX$ is the homomorphism induced by the operation $G \to \operatorname{Aut} X$. In the upper right, we cannot write DX, because only the complete vectorfields are sent into $\operatorname{Aut} X$ by the exponential map. $\exp$ is therefore, even in this case, a natural transformation (of suitably defined functors).

6.2. <u>exp is a local diffeomorphism at the identity.</u> We show now

PROPOSITION 6.2.1. <u>The tangent linear map</u> $\exp_{*_0}: G_e \to G_e$ <u>induced by</u> $\exp: G_e \to G$ <u>is the identity map.</u>

<u>Proof:</u> For $A \in G_e$ we have

$$\exp_{*_0} A = \exp_{*_{tA}} A \Big|_{t=0} = \{\exp_{*_{tA}} \frac{d}{dt}(tA)\}\Big|_{t=0}$$

$$= \frac{d}{dt} \exp(tA)\Big|_{t=0} = A \,, \qquad \text{q.e.d.}$$

By the inverse function theorem we therefore have

THEOREM 6.2.2. <u>There is an open neighborhood</u> N_0 <u>of</u> O <u>in</u> G_e <u>and an open neighborhood</u> N_e <u>of</u> e <u>in</u> G <u>such that</u> $\exp: N_0 \to N_e$ <u>is an analytic diffeomorphism.</u>

We denote by $\log: N_e \to N_0$ the inverse map. The map log defines a chart of G at e.

DEFINITION 6.2.3. <u>A canonical chart of G</u> is a pair $(N_e, \log)$ of an open neighborhood N_e of e in G and a diffeomorphism $\log: N_e \to \log(N_e) = N_0 \hookrightarrow G_e$ which is an inverse of $\exp/N_0$.

An immediate application of theorem 6.2.2 is the following

PROPOSITION 6.2.4. *Let* G *be a Lie group and* G_0 *the connected component of the identity. If* LG *is commutative, then* G_0 *is commutative.*

Proof: The image of exp contains an open neighborhood U of e in G. Any two elements in U commute, as $\exp: G_e \to G$ is a homomorphism by proposition 6.1.4. But U generates G_0, so that any two elements of G_0 commute.

Together with corollary 4.6.9 we have therefore

THEOREM 6.2.5. *Let* G *be a connected Lie group. Then* G *is commutative if and only if* LG *is commutative.*

LEMMA 6.2.6. *Let* $\rho: G \to G'$ *be a homomorphism.* $L(\rho): LG \to LG'$ *is injective (surjective) if and only if* ρ_{*_g} *is injective (surjective) for every* $g \in G$.

Proof: $\rho(g\gamma) = \rho(g)\rho(\gamma)$ implies for $\gamma = e$ $\rho_{*_g} \circ (Lg)_{*_e}$ $= (L_{\rho(g)})_{*_{e'}} \circ \rho_{*_e}$ and $\rho_{*_g} = (L_{\rho(g)})_{*_{e'}} \circ \rho_{*_e} \circ (L_g)_{*_e}^{-1}$, q.e.d.

PROPOSITION 6.2.7. *Let* G *be a commutative connected Lie group. Then* $\exp: LG \to G$ *is surjective.*

Proof: We have seen in 6.1.4 that for commutative LG $\exp: LG \to G$ is a homomorphism. Let G' be the image. Now

exp_{*0} is the identity map and therefore exp_{*A}, by 6.2.6, an isomorphism for every $A \in LG$. Therefore G' is an open (and closed) subgroup of G, i.e. $G' = G$.

Consider a local homomorphism $\rho: G \to G'$. Let $(N_e, \log)$ be a canonical chart of G. By the naturality 6.1.6 of exp, we have for $g \in N_e$

$$(*) \qquad \rho(g) = \exp(L(\rho)\log g)$$

This necessary condition determines ρ/N_e by $L(\rho)$ and has the following applications.

PROPOSITION 6.2.8. <u>Let</u> $\rho_i: G \to G'$ $(i = 1, 2)$ <u>be local homomorphisms. If the induced algebra homomorphisms</u> $L(\rho_i): LG \to LG'$ $(i = 1, 2)$ <u>coincide, then there exists an open neighborhood</u> U <u>of</u> e <u>in</u> G, <u>on which</u> ρ_1 <u>and</u> ρ_2 <u>coincide.</u>

<u>Proof:</u> Take $U = N_e$ as the domain of a canonical chart of G. Then $(*)$ shows $\rho_1(g) = \rho_2(g)$ for $g \in U$.

COROLLARY 6.2.9. <u>Let</u> G <u>be connected and</u> $\rho_i: G \to G'$ $(i = 1, 2)$ <u>homomorphisms.</u> <u>If</u> $L(\rho_1) = L(\rho_2)$, <u>then</u> $\rho_1 = \rho_2$.

<u>Proof:</u> Any neighborhood of e generates G. ▮

This corollary of 6.2.8 is expressed by saying that the functor $L: \mathfrak{Lg} \to \mathfrak{LA}$ is faithful on the subcategory of connected Lie groups and global homomorphisms.

An application of this last result is the

PROPOSITION 6.2.10. *Let G be connected. Then* $\ker \mathrm{Ad} = ZG$, *where* ZG *is the center of* G.

Proof: By definition $\mathrm{Ad} = L \circ \mathfrak{J}$. But as seen before, $L(\mathfrak{J}_g) = 1_{LG}$ implies $\mathfrak{J}_g = 1_G$. Therefore $\ker \mathrm{Ad} = \ker \mathfrak{J} = ZG$, q.e.d.

Now consider the problem of constructing a local homomorphism $G \to G'$ inducing a given Lie algebra homomorphism $LG \to LG'$. Remember that the isomorphism $L\mathbb{T} \to L\mathbb{R}$ is not induced by any homomorphism $\mathbb{T} \to \mathbb{R}$, so we only can expect the existence of a local homomorphism with the desired property.

PROPOSITION 6.2.11. *Let* G, G' *be Lie groups and* $h: LG \to LG'$ *a Lie algebra homomorphism. Let* $(N_e, \log)$ *be a canonical chart at* e *in* G. *The restriction to* N_e *of a local homomorphism* $G \to G'$ *inducing* h *is necessarily of the form*

$$\rho = \exp \circ h \circ \log: N_e \longrightarrow G'$$

If LG *and* LG' *are commutative, then* $\rho: N_e \to G'$ *defined by this formula is a local homomorphism inducing* h.

Proof: We have already seen that the restriction of a local homomorphism $G \to G'$ to N_e is necessarily of this form. By 6.1.4 for commutative LG exp is a homomorphism $G_e \to G$, so that log is also a local homomorphism $G \to G_e$. Therefore ρ is a homomorphism. Let $L(\rho): LG \to LG'$ be the induced homomorphism.

It is clear that $\rho_{*_e}: G_e \to G_e$ is just h. This shows $L(\rho) = h$. ∎

The map $\rho: N_e \to G'$ defined by a Lie algebra homomorphism $h: LG \to LG'$ is a local homomorphism $G \to G'$ inducing h even without supposing LG and LG' commutative. But a direct proof of this requires a deeper analysis of the situation. See also the comments in section 6.4, after proposition 6.4.2. We shall construct in Chapter 7, by a different method, a local homomorphism $G \to G'$ inducing a given Lie algebra homomorphism $h: LG \to LG'$ (see 7.2.3). By unicity, this will prove that the $\rho: N_e \to G'$ defined in proposition 6.2.11 is a local homomorphism.

Exercise 6.2.12. Suppose that a Lie group G operates effectively on the manifold X with respect to $\tau: G \to \mathrm{Aut}\, X$, and let KX be the Lie algebra of Killing vectorfields on X. Show that $g \in G$ satisfies $(\tau_g)_* A = A$ for every $A \in KX$ if and only if g is in the centralizer of the identity component G_0 in G.

6.3. Unicity of Lie group structure. We begin by proving the following important

PROPOSITION 6.3.1. *Let G be a Lie group and $\alpha: \mathbb{R} \to G$ a homomorphism in the algebraic sense, which is continuous. Then there exists an $A \in LG$, such that $\alpha_t = \exp tA$ and hence α is analytic, i.e. a 1-parameter subgroup of G.*

Proof: Let $(U, \log)$ be a canonical chart of G and V a neighborhood of e in G with $VV \subset U$.

Let $g \in V$. Then $g^2 \in VV \subset U$ and $\log g$, $\log g^2$ are defined. Consider the 1-parameter subgroup of G $t \rightsquigarrow \exp(t \log g)$. For $t = 1$ in particular $\exp \log g = g$. g^2 is on this 1-parameter subgroup and $g^2 = \exp(2 \log g)$. On the other hand, $g^2 = \exp \log g^2$, as $g^2 \in U$. Therefore $\log g^2 = 2 \log g$ or $g = \exp(\frac{1}{2} \log g^2)$, which means that g is uniquely determined by g^2.

Now consider the continuous homomorphism $\alpha: \mathbb{R} \to G$. There exists an $\epsilon > 0$ such that $\alpha_t \in V$ for every t with $|t| \leqq \epsilon$. We can suppose $\epsilon = 1$ (otherwise change the parameter t by λt such that the new parameter is defined for absolute values $\leqq 1$). Define $\alpha_1 = g \in V$. Now $\exp(\frac{1}{2} \log g)$ is a square root of g in V, by the preceding therefore the unique one. This shows $\alpha_{1/2} = \exp(\frac{1}{2} \log g)$, or with $\log \alpha_1 = A$ also $\log \alpha_{1/2} = \frac{1}{2} A$. By iteration one obtains $\log \alpha_{(1/2^n)} = \frac{1}{2^n} A$ and by addition

$$\log \alpha_{\left(\frac{p}{2^n}\right)} = \frac{p}{2^n} A \qquad \text{for } 0 \leqq p \leqq 2^n, \; p \in \mathbb{N}^* .$$

This shows $\log \alpha_r = rA$ for every dyadic rational r with $0 \leqq r \leqq 1$ and by continuity $\log \alpha_t = tA$. This proves $\alpha_t = \exp tA$. ∎

To generalize 6.3.1 to arbitrary homomorphisms, we shall make use of

LEMMA 6.3.2. Let G be a Lie group. Suppose G_e is a direct product $M \times N$ of vector subspaces M, N. Then the map $\phi: M \times N \to G$ defined by $\phi(A, B) = \exp A \exp B$ for $A \in M$, $B \in N$ is a local diffeomorphism at 0.

<u>Proof:</u> In view of the inverse function theorem, we only have to show that $\phi_{*_0}: M \times N \to G_e$ is an isomorphism. Now $\phi = m \circ (\exp/M \times \exp/N)$, where m denotes the multiplication $m: G \times G \to G$. Therefore for $(X, Y) \in M \times N$ we have

$$\phi_{*_0}(X, Y) = m_{*_{(e,e)}}(\exp_{*_0} X, \exp_{*_0} Y) = \exp_{*_0} X + \exp_{*_0} Y$$

$$= X + Y ,$$

as $\exp_{*_0}$ = identity by 6.2.1. Hence ϕ_{*_0} is the identity and the lemma is proved.

Remark. The lemma generalizes of course to the case of a decomposition $G_e = M_1 \times \ldots \times M_n$ for a finite number of vector-spaces $M_i \subset G_e$.

We also state the following

LEMMA 6.3.3. <u>Let</u> G, G' <u>be Lie groups and</u> $\rho: G \to G'$ <u>a homomorphism in the algebraic sense.</u> <u>If</u> ρ <u>is differentiable (analytic) at</u> e, <u>then</u> ρ <u>is everywhere differentiable (analytic).</u>

<u>Proof:</u> Clear from $\rho \circ L_g = L_{\rho(g)} \circ \rho$. ∎

We are now able to prove

THEOREM 6.3.4. <u>Let</u> G, G' <u>be Lie groups and</u> $\rho: G \to G'$ <u>a homomorphism of groups in the algebraic sense, which is continuous. Then</u> ρ <u>is analytic, i.e. a homomorphism of Lie groups.</u>

Proof: Let $A \in G_e$. The correspondence $t \rightsquigarrow \rho(\exp tA)$ is a continuous homomorphism $\mathbb{R} \to G$. Hence there exists an $A' \in G'_e$, such that $\rho(\exp tA) = \exp tA'$.

Now let A_i $(i = 1, \ldots, n = \dim G)$ be a base of G_e, and define A'_i as the vector in G'_e with $\rho(\exp tA_i) = \exp tA'_i$. Then $\rho(\Pi_{i=1}^n \exp t_i A_i) = \Pi_{i=1}^n \exp t_i A'_i$.

Now the map $\phi: \mathbb{R}^n \to G$ defined by $\phi(t_1, \ldots, t_n) = \Pi_{i=1}^n \exp t_i A_i$ is a local diffeomorphism at 0 by the remark following 6.3.2. Therefore there exists a neighborhood V of e in G such that every $g \in V$ may be written in the form $g = \Pi_{i=1}^n \exp t_i A_i$, with t_i depending analytically of g. The formula $\rho(\Pi_{i=1}^n \exp t_i A_i) = \Pi_{i=1}^n \exp t_i A'_i$ now shows that ρ is analytic at the neutral element. Therefore ρ is analytic by 6.3.3.

Remark. The map $\rho_{*_e}: G_e \to G'_e$ is just given by $\rho_{*_e} A_i = A'_i$ for $i = 1, \ldots, n$.

COROLLARY 6.3.5. Let G, G' be Lie groups. If $G = G'$ as topological groups, then $G = G'$ as Lie groups.

Proof: If $G = G'$ as topological groups, the identity map is a homeomorphism, and therefore a diffeomorphism by 6.3.4.

This shows that the Lie algebra of a Lie group is in fact a property of the underlying topological group.

This raises the question: which topological groups can be turned into Lie groups, i.e. have an analytic structure compatible with the group structure and such that the corresponding topology coincides with the given one ?

It has been proved by A. M. Gleason, Ann. of Math. 56 (1952), 193-212, that a topological group G which is locally compact, locally connected, metrisable and of finite dimension, is a Lie group.

6.4. Application to fixed points on G-manifolds. As an application we give, in this section, a characterization of fixed points on a G-manifold by the Lie algebra of Killing vectorfields. We begin with the following

LEMMA 6.4.1. Let X be a manifold, A a vectorfield and φ_t a local 1-parameter group of local transformations generated by A. A point $x \in X$ is a fixpoint of every transformation φ_t if and only if $A_x = 0$.

Proof: $\varphi_t(x) = x$ for every t implies $\dot{\varphi}_t(x)|_{t=0} = 0$ and therefore $A_x = 0$. Conversely, let $A_x = 0$. Then the differential equation $\dot{\varphi}_t(x) = A_{\varphi_t(x)}$ has the solution $\varphi_t(x) = x$ for every t, and the solution is unique.

Example 6.4.2. On the two-sphere S^2 every vectorfield has a zero. Therefore every 1-parameter group of transformations has a fixpoint.

More generally, let X be a compact manifold. The vanishing of the Euler-Poincaré characteristic $\chi(X)$ is necessary and sufficient for the existence of a vectorfield without zeros. (Remember that vectorfield means differentiable vectorfield). Therefore every 1-parameter group of transformations of a compact manifold X with $\chi(X) \neq 0$ has a fixpoint.

PROPOSITION 6.4.3. <u>Let the connected Lie group</u> G <u>operate on</u> X <u>by</u> $\tau: G \to \mathrm{Aut}\, X$ <u>and let</u> KX <u>be the Lie algebra of Killing vectorfields on</u> X. <u>A point</u> $x \in X$ <u>is</u> G-<u>invariant if and only if</u> $A^*_x = 0$ <u>for every</u> $A^* \in KX$.

<u>Proof</u>: Suppose x G-invariant. For any $a \in \mathfrak{L}G$ we have $\tau_{a_t}(x) = x$ and therefore $A^*_x = 0$ for the corresponding Killing vectorfield on X. Suppose conversely $A^*_x = 0$ for every $A^* \in KX$. By lemma 6.4.1, for every $a \in \mathfrak{L}G$ we have therefore $\tau_{a_t}(x) = x$. exp being a local diffeomorphism, there is an open neighborhood U of e in G such that $\tau_g(x) = x$ for $g \in U$. As G is connected, U generates G and $\tau_g(x) = x$ for every $g \in G$. ∎

Consider in particular a finite-dimensional $\mathbb{R}$-vectorspace V and a representation $\tau: G \to GL(V)$ of the connected Lie group G in V. A point $v \in V$ is G-invariant if and only if $A^*_v = 0$ for every $A^* \in KV$. But, by example 5.5.5 the Killing vectorfield A^* corresponding to $a \in \mathfrak{L}G$ is defined by $A^*_v = \tau_{*_e} \dot{a}_0 v$. Therefore $v \in V$ is G-invariant if and only if $(\tau_{*_e} A)v = 0$ for every $A \in G_e$. Considering the induced representation $L(\tau): LG \to \mathfrak{L}(V)$ of LG in V, we see that $v \in V$ is G-invariant if and only if $(L(\tau)A)v = 0$ for every $A \in LG$.

This motivates the following

DEFINITION 6.4.4. Let Λ be field, M a Λ-Lie algebra, V a Λ-vectorspace and $\sigma: M \to \mathfrak{L}(V)$ a representation of M in V. An element $v \in V$ is called invariant or M-invariant if $\sigma(A)v = 0$ for every $A \in M$.

We have therefore

PROPOSITION 6.4.5. Let V be a finite dimensional $\mathbb{R}$-vectorspace, G a connected Lie group, $\tau: G \to GL(V)$ a representation of G in V and $L(\tau): LG \to \mathfrak{L}(V)$ the induced representation of LG in V. An element $v \in V$ is G-invariant if and only if it is LG-invariant.

We apply now proposition 6.4.3 for the case of a commutative G and prove

PROPOSITION 6.4.6. Let X be a manifold. The following conditions are equivalent.

(1) For any n-dimensional commutative, connected Lie group G and any operation $\tau: G \to \mathrm{Aut}\, X$ there is a G-invariant point $x \in X$.

(2) Any operation of the additive group $\mathbb{R}^n$ on X has a fixpoint.

(3) For any n-tuple $A_1, \ldots, A_n$ of complete vectorfields with $[A_i, A_j] = 0$ $(i, j = 1, \ldots, n)$ there exists a point $x \in X$ with $A_{i_x} = 0$ for $i = 1, \ldots, n$.

Proof: (1) $\Rightarrow$ (2) is clear.

(2) $\Rightarrow$ (3) Let $A_1, \ldots, A_n$ be complete vectorfields on X with $[A_i, A_j] = 0$ for $i, j = 1, \ldots, n$. Consider the additive group K generated by $A_1, \ldots, A_n$. By proposition 5.6.6 there is an operation of K on X, such that the commutative Lie algebra K is the algebra of Killing vectorfields of this operation. K is isomorphic to the additive group $\mathbb{R}^k$ for some $k \leqq n$. The operation of K on X induces therefore an operation of the additive group $\mathbb{R}^n$ on X, which has a fixpoint x by hypothesis. Now by proposition 6.4.3, x is a zero for any $A \in K$ and in particular $A_{i_x} = 0$ for $i = 1, \ldots, n$.

(3) $\Rightarrow$ (1). Let G be a n-dimensional commutative, connected Lie group G, $\tau: G \to \mathrm{Aut}\, X$ an operation of G on X and KX the Lie algebra of Killing vectorfields. We have $\dim KX \leqq n$. Let $A_1, \ldots, A_n$ be a system of generators of KX. Then $[A_i, A_j] = 0$ for $i, j = 1, \ldots, n$. By hypothesis, there exists a common zero $x \in X$ for these vectorfields. Then x is a zero for any Killing vectorfield and by proposition 6.4.3 x is G-invariant. ∎

If one (and hence any) of the conditions in proposition 6.4.6 is satisfied for a certain n, then it is clearly also satisfied for every $m \leqq n$.

The underlying manifold of a commutative Lie group G of dimension n clearly doesn't satisfy any of the conditions of the proposition 6.4.6. The operation of G_0 on G by translations has no fixpoints and any n-tuple of invariant vectorfields $A_1, \ldots, A_n$

satisfies $[A_i, A_j] = 0$ without any one of the vectorfields having a zero.

Example 6.4.7. Consider the two-sphere S^2. For $n=1$, the condition (3) of proposition 6.4.6 just says that every vectorfield on S^2 has a zero, which is a consequence of $\chi(S^2) \neq 0$. For $n = 2$, the condition (2) was shown to be satisfied by E. L. Lima, Proc. AMS, Vol. 15 (1964), p. 138-141.

6.5. Taylor's formula. In this section we make essential use of the analycity of G. We recall that in this chapter we identify LG with G_e.

PROPOSITION 6.5.1. Let $f \in CG$ be a function analytic at $g \in G$, and $A \in LG$. Then there exists an $\epsilon > 0$ such that

$$f(g \exp tA) = \sum_{n=0}^{\infty} \frac{t^n}{n!} [A^n f](g) \qquad \text{for } |t| < \epsilon .$$

Proof: First, let $f \in CG$. By proposition 5.4.7

$$[Af](g) = \frac{d}{dt} f(g \exp tA) \Big|_{t=0}$$

This proves

$$(*) \qquad [A^n f](g) = \left[\left(\frac{d}{dt}\right)^n f(g \exp tA) \right]_{t=0}$$

for $n = 1$. We prove (*) for arbitrary n by induction.

$$[A^{n+1}f](g) = [A^n(Af)](g) = \left[\left(\frac{d}{dt}\right)^n (Af)(g \exp tA)\right]_{t=0}$$

$$= \left[\left(\frac{d}{dt}\right)^n \frac{d}{dn} f(g \exp tA \exp uA)\right]_{\substack{t=0\\u=0}}$$

$$= \left[\left(\frac{d}{dv}\right)^n \frac{d}{dv} f(g \exp vA)\right]_{v=0}$$

with $t + u = v$, showing thus (*).

If f is now analytic at g, then there exists an $\epsilon > 0$ such that for $|t| < \epsilon$

$$f(g \exp tA) = \sum_{n=0}^{\infty} \frac{t^n}{n!} \left[\left(\frac{d}{dt}\right)^n f(g \exp tA)\right]_{t=0}$$

$$= \sum_{n=0}^{\infty} \frac{t^n}{n!} \left[A^n f\right](g) \qquad \text{, q. e. d.}$$

We apply this to prove the

PROPOSITION 6.5.2. Let $O(t^3)$ denote a vector in LG such that for an $\epsilon > 0$ $\frac{1}{t^3} O(t^3)$ is bounded and analytic for $|t| < \epsilon$. Then for $A, B \in LG$ and sufficiently small t

(i) $\exp tA \exp tB = \exp \{t(A+B) + \frac{t^2}{2} [A, B] + O(t^3)\}$

(ii) $\exp tA \exp tB \exp(-tA) = \exp \{tB + t^2[A, B] + O(t^3)\}$

(iii) $\exp(-tA) \exp(-tB) \exp tA \exp tB = \exp \{t^2[A, B] + O(t^3)\}$.

Proof: Let f be analytic at e. We have shown

$$[A^n f](e) = \left[\left(\frac{d}{dt}\right)^n f(\exp tA)\right]_{t=0} \quad .$$

We obtain therefore

$$[A^nB^mf](e) = \left[\left(\frac{d}{dt}\right)^n\left(\frac{d}{ds}\right)^m f(\exp tA \exp sB)\right]_{\substack{t=0\\ s=0}}$$

The Taylor series for $f(\exp tA \exp sB)$ is therefore

$$f(\exp tA \exp sB) = \sum_{n,m \geqq 0} \frac{t^n}{n!}\frac{s^m}{m!}[A^nB^mf](e)$$

and for $t = s$

(1) $$f(\exp tA \exp tB) = \sum_{n,m \geqq 0} \frac{t^{n+m}}{n!\,m!}[A^nB^mf](e)$$

The coefficient of t is $\{[Af](e) + [Bf](e)\}$,

the coefficient of t^2 is $\{\frac{1}{2}[A^2f](e) + [ABf](e) + \frac{1}{2}[B^2f](e)\}$.

On the other hand, by theorem 6.2.2 for sufficiently small t

$$\exp tA \exp tB = \exp Z(t)$$

with $Z: I \to G_e$, I an open interval of $\mathbb{R}$ containing 0 , Z analytic at 0 , $Z(0) = 0$. Then

$$Z(t) = tZ_1 + t^2Z_2 + O(t^3)$$

for fixed $Z_1, Z_2 \in G_e$.

Take any function f which is linear in a canonical chart at e . Then it is analytic at e and

$$(2) \qquad f(\exp tA \exp tB) = f(\exp\{tZ_1 + t^2 Z_2 + O(t^3)\})$$
$$= f(\exp\{tZ_1 + t^2 Z_2\}) + O'(t^3)$$
$$= \sum_{n=0}^{\infty} \frac{1}{n!} [(tZ_1 + t^2 Z_2)^n f](e) + O'(t^3)$$

$O'(t)$ being a real number such that for an $\epsilon > 0$ $\frac{1}{t^3} O'(t)$ is bounded and analytic for $|t| < \epsilon$.

The coefficient of t is $[Z_1 f](e)$, the coefficient of t^2 is $\{[Z_2 f](e) + \frac{1}{2}[Z_1^2 f](e)\}$.

Comparing this with the coefficient of t and t^2 in (1), we obtain

$$[Z_1 f](e) = \{(A+B)f\}(e)$$
$$[Z_2 f](e) = \{\tfrac{1}{2}[A,B]f\}(e) \quad .$$

This being true for any function f which is linear in a canonical chart at e, we have therefore

$$Z_1 = A + B$$
$$Z_2 = \tfrac{1}{2}[A,B]$$

This shows

$$\exp tA \exp tB = \exp Z(t) = \exp\{t(A+B) + \frac{t^2}{2}[A,B] + O(t^3)\} ,$$

i.e. (i) .

(ii) is obtained by (i) as follows

$$\begin{aligned}\exp tA \exp tB \exp(-tA) &= \exp(t\{(A+B)+\tfrac{t}{2}[A,B]+O(t^2)\}_1)\exp(-tA)\\ &= \exp(t(\{\ \}_1-A)+\tfrac{t^2}{2}[\{\ \}_1,-A]+O(t^3))\\ &= \exp((tB+\tfrac{t^2}{2}[A,B])+\tfrac{t^2}{2}[A,B]+O(t^3))\\ &= \exp(tB+t^2[A,B]+O(t^3))\end{aligned}$$

(iii) is shown similarly by

$$\begin{aligned}\exp(-tA)\exp(-tB)\exp tA \exp tB &= \exp(t\{-(A+B)+\tfrac{t}{2}[A,B]+O(t^2)\}_2)\cdot\\ &\quad \exp(t\{(A+B)+\tfrac{t}{2}[A,B]+O(t^2)\}_1)\\ &= \exp(t(\{\ \}_2+\{\ \}_1)+\tfrac{t^2}{2}[\{\ \}_2,\{\ \}_1]+O(\\ &= \exp(t^2[A,B]+O(t^3)) \quad , \text{ q.e.d.}\end{aligned}$$

Remark. Let N_0 be an open neighborhood of O in G_e such that the restriction $\exp/N_0 : N_0 \to N_e$ is a diffeomorphism. Then one can define a composition

$$A \circ B = \log(\exp A \cdot \exp B) \qquad \text{for } A, B \in N_0$$

if $\exp A \cdot \exp B \in N_e$. This defines a (partial) composition law in N_0 for which O is an identity. In fact, by the very definition of this composition, exp is an isomorphism of N_0 with N_e equipped with the composition inherited from G. Now look at the formula (i) of proposition 6.5.2. It can be rewritten (for arbitrary $A, B \in LG$ and sufficiently small t) as

$$tA \circ tB = (tA + tB) + \frac{1}{2}[tA, tB] + O(t^3) \quad .$$

The fundamental fact can be proved that (for sufficiently small t) the term $O(t^3)$ also is expressable by operations in LG on A, B. This means that the composition law in the neighborhood N_e of e in G is completely determined by the Lie algebra LG. The formula (i) of 6.5.2 just gives the first two terms of this development.

Moreover one can show that a Lie algebra homomorphism $h: LG \to LG'$ is a homomorphism with respect to the composition defined in N_0. This incidentally shows that the map $\rho: N_e \to G'$ determined by $h: LG \to LG'$ according to proposition 6.2.9 is in fact a local homomorphism $G \to G'$ inducing $h: LG \to LG'$.

We apply proposition 6.5.2 to prove

PROPOSITION 6.5.3. Let $A, B \in LG$. Then the following conditions are equivalent

(i) $[A, B] = 0$

(ii) $\exp sA \exp tB = \exp tB \exp sA$ for every s and t

(iii) $\exp tA \exp tB = \exp tB \exp tA$ for every t.

Proof: (i) ⇒ (ii) by proposition 5.4.11. (ii) ⇒ (iii) is trivial. To see (iii) ⇒ (i) observe that (iii) implies by proposition 6.5.2

$$\exp\{t(A+B) + \frac{t^2}{2}[A, B] + O(t^3)\} = \exp\{t(B+A) + \frac{t^2}{2}[B, A] + O(t^3)\}$$

for sufficiently small t. This implies $[A, B] = [B, A]$ and $[A, B] = 0$.

The condition (i) and (iii) of proposition 6.5.2 also imply the following convenient formulas.

COROLLARY 6.5.4. <u>Under the conditions of proposition 6.5.2, one has</u>

(i) $\exp t(A+B) = \exp tA \exp tB \exp\left\{\frac{t^2}{2}[A,B] + O(t^3)\right\} = \exp tA \exp tB \exp O(t^2)$

(ii) $\exp\{t^2[A,B]\} = \exp(-tA)\exp(-tB)\exp tA \exp tB \exp O(t^3)$.

<u>Proof:</u> (i) follows from

$$\exp(-tA)\exp(-tB)\exp t(A+B) = \exp(t\{-(A+B) + \frac{t}{2}[A,B] + O(t^2)\})\exp t(A+B)$$

$$= \exp(t(\{\ \} + (A+B)) + \frac{t^2}{2}[\{\ \}, A+B] + O(t^3))$$

$$= \exp(\frac{t^2}{2}[A,B] + O(t^3)) .$$

(ii) follows from

$$\exp(-tB)\exp(-tA)\exp tB \exp tA \exp(t^2[A,B]) = \exp\{t^2[B,A] + O(t^3)\}\exp\{t^2[A,B]\}$$

$$= \exp O(t^3) .$$

The formula (i) shows that the curve $t \rightsquigarrow \exp tA \exp tB$ has the same tangent vector at e than the 1-parameter subgroup $t \rightsquigarrow \exp t(A+B)$. From proposition 5.4.10 it follows that the term $O(t^3)$ is vanishing for A, B with $[A,B] = 0$.

The formula (ii) describes $[A,B]$ as the tangent vector at e of the curve $t \rightsquigarrow \exp(-\sqrt{t}A)\exp(-\sqrt{t}B)\exp\sqrt{t}A \exp\sqrt{t}B$.

Another consequence of 6.5.2 useful for later application is the following

COROLLARY 6.5.5. <u>Let</u> $A, B \in LG$. <u>Then for any</u> $t \in \mathbb{R}$ <u>we have</u>

(i) $\exp t(A+B) = \lim_{n\to\infty} \left\{\exp \frac{t}{n} A \exp \frac{t}{n} B\right\}^n$

(ii) $\exp\{t^2[A,B]\} = \lim_{n\to\infty} \left\{\exp\left(-\frac{t}{n}A\right) \exp\left(-\frac{t}{n}B\right) \exp \frac{t}{n} A \exp \frac{t}{n} B\right\}^{n^2}$

Proof: Let $t \in \mathbb{R}$ and n sufficiently great. By proposition 6.5.2 for fixed t

$$\exp \frac{t}{n} A \exp \frac{t}{n} B = \exp\left\{\frac{t}{n}(A+B) + \frac{t^2}{2n^2}[A,B] + O\left(\frac{1}{n^3}\right)\right\}$$

and therefore

$$\left\{\exp \frac{t}{n} A \exp \frac{t}{n} B\right\}^n = \exp\left\{t(A+B) + \frac{t^2}{2n}[A,B] + O\left(\frac{1}{n^2}\right)\right\}$$

thus showing (i). To see (ii) it suffices to observe that by 6.5.2

$$\left\{\exp\left(-\frac{t}{n}A\right) \exp\left(-\frac{t}{n}B\right) \exp \frac{t}{n} A \exp \frac{t}{n} B\right\}^{n^2} = \left\{\exp\left\{\frac{t^2}{n^2}[A,B] + O\left(\frac{1}{n^3}\right)\right\}\right\}^{n^2}$$

$$= \exp\{t^2[A,B] + O(\tfrac{1}{n})\} .$$

CHAPTER 7. SUBGROUPS AND SUBALGEBRAS

7.1. Lie subgroups. Before defining the notion of Lie subgroups, we prove

LEMMA 7.1.1. *Let* H, G *be Lie groups and* $\iota: H \to G$ *an injective homomorphism. Then the induced homomorphism* $L(\iota): LH \to LG$ *is injective.*

Proof: Let $\alpha_i \in \mathfrak{L}H (i = 1, 2)$ be 1-parameter subgroups of H with $\iota \circ \alpha_1 = \iota \circ \alpha_2$. ι being injective, $\alpha_1 = \alpha_2$. Therefore the map $\mathfrak{L}(\iota): \mathfrak{L}H \to \mathfrak{L}G$ induced by ι is injective. By 5.4.12, this shows the injectivity of $L(\iota)$, q.e.d.

Note that by lemma 6.2.6 every tangent linear map $\iota_{*h}: H_h \to G_{\iota(h)}$ is injective.

DEFINITION 7.1.2. Let G be a Lie group. A subgroup H of G is a Lie subgroup of G if

(i) H is a Lie group

(ii) the injection $\iota: H \hookrightarrow G$ is analytic.

Let H be a Lie subgroup of G. By lemma 7.1.1 and the remark following it, the pair (H, ι) is a submanifold of G according to the

DEFINITION 7.1.3. Let G be a manifold. A subset H of G is a submanifold of G if

(i) H is a manifold

(ii) the injection $\iota: H \hookrightarrow G$ is an immersion, i.e. ι differentiable and $\iota_{*h}: H_h \to G_{\iota(h)}$ injective for any $h \in H$.

Let H be a subgroup of G. If H is a submanifold of G, locally the analytic structure of H is induced from that of G, so that the group operations in H are analytic. H is therefore a Lie subgroup of G.

Note that a 1-parameter subgroup $\alpha: \mathbb{R} \to G$ is a Lie subgroup if and only if α is injective.

Let H be a Lie subgroup of G. By 7.1.1, the injection $\iota: H \hookrightarrow G$ induces an injective map $L(\iota): LH \to LG$. We can therefore identify LH with a subalgebra of LG and write $L(\iota): LH \hookrightarrow LG$.

LEMMA 7.1.4. Let H be a Lie subgroup of G. The map $\exp: H_e \to H$ is the restriction of $\exp: G_e \to G$.

Proof: After the canonical identifications, this is just the naturality 6.1.6 of the exponential map.

PROPOSITION 7.1.5. Let H_i $(i = 1, 2)$ be connected Lie subgroups of G. If $LH_1 = LH_2$, then $H_1 = H_2$.

Proof: There is an open neighborhood of e in H_1 which is also an open neighborhood of e in H_2 (take a canonical chart at e and use 7.1.4.).

We use without proof the following

LEMMA 7.1.6. Let X, Y be manifolds, S a submanifold of Y and $\varphi: X \to Y$ a differentiable map with $\varphi(X) \subset S$. If the induced map $\hat{\varphi}: X \to S$ is continuous, it is differentiable.

LEMMA 7.1.7. Let G be a Lie group and H a Lie subgroup. Then $LH = \{A \in LG / t \leadsto \exp tA$ is a continuous map $\mathbb{R} \to H\}$.

Proof: $A \in LH$ implies that $t \leadsto \exp tA$ is a differentiable map $\mathbb{R} \to H$. Suppose conversely $A \in LG$ with $t \leadsto \exp tA$ a continuous map $\mathbb{R} \to H$. This map is differentiable by lemma 7.1.6 and therefore $A \in LH$.

PROPOSITION 7.1.8. Let H_1, H_2 be two Lie subgroups of G . If H_1 and H_2 coincide as topological groups, they coincide as Lie groups.

Proof: 7.1.7 characterizes the Lie algebra by aid of the topological structure alone. By 7.1.5, $H_{1_0} = H_{2_0}$. The identity map $H_1 \to H_2$ is therefore an isomorphism.

This is of course also a consequence of 6.3.5, but we have preferred a simple, direct proof.

We state now

THEOREM 7.1.9. Let G be a Lie group. If H is a Lie subgroup of G , then the Lie algebra of H is a subalgebra of LG . Each subalgebra of LG is the Lie algebra of a unique connected Lie subgroup of G .

Proof: There only remains to show, that for a given subalgebra $\mathfrak{h}$ of LG there exists a connected Lie subgroup H of G with $LH = \mathfrak{h}$.

Suppose there exists such a Lie subgroup H. Then $\exp(\mathfrak{h}) \subset H$. Moreover $\exp \mathfrak{h}$ contains an open neighborhood of e in H and therefore generates H.

This permits conversely to define H as the subgroup generated by $\exp \mathfrak{h}$. The problem is to make H a submanifold of G. We do not show this here, but sketch instead another proof (see Chevalley [3], p. 109, theorem 1), making use of the existence theorem for integral manifolds of an involutive field of vectorspaces on a manifold.

Let H be a Lie subgroup of G. Then the left cosets of G modulo H are the maximal integral manifolds of the field of vectorspaces W on G defined by the tangentspaces of the cosets. Now given conversely a subalgebra $\mathfrak{h} \subset LG$ one can reconstruct the field of vectorspaces W. W_g is namely the vectorspace $\{A_g / A \in \mathfrak{h}\}$. $\mathfrak{h}$ being a subalgebra, W is then involutive. Let H be the maximal integral manifold of W passing through e. To see that H is a subgroup of G, first observe that the field of vectorspaces W is invariant by left translations. Therefore the maximal integral manifolds are just permuted among themselves by left translations. Now if $h \in H$, then $L_{h^{-1}} h = e$, so $L_{h^{-1}} H = H$. If conversely $L_{g^{-1}} H = H$ for $g \in G$, then $g \in H$. Therefore $H = \{g / L_{g^{-1}} H = H\}$ and H is a subgroup of G. H being a submanifold of G, we see that H is a Lie subgroup of G.

Exercise 7.1.10. Let X be a G-manifold and H a Lie subgroup of G. Then X is a H-manifold. Suppose $\tau: G \to \operatorname{Aut} X$

to be an effective operation of G on X and let S be a subalgebra of the Lie algebra KX of Killingvectorfields. Then there is a unique connected Lie subgroup H of G such that the restriction of τ to H defines an operation on X with S as Lie algebra of Killing-vectorfields.

7.2. Existence of local homomorphisms. We begin with

LEMMA 7.2.1. *Let* $\rho: G \to G'$ *be a local homomorphism of Lie groups. If* $L(\rho): LG \to LG'$ *is an isomorphism, then* ρ *is a local isomorphism.*

Proof: If $L(\rho)$ is an isomorphism, there exists on an open neighborhood of e' in G a local inverse map μ of $\rho: G \to G'$. ρ being a local homomorphism, μ is necessarily a local homomorphism and ρ therefore a local isomorphism. ∎

Example 7.2.2. If G is commutative, $\exp: LG \to G$ is a homomorphism by 6.1.4. Now $L(\exp) = 1_{LG}: LG \to LG$ and $\exp$ is moreover surjective by 6.2.7. This is sufficient to determine the structure of commutative connected Lie groups (see section 7.3).

We have already proved in 6.2.11 the existence of a local homomorphism $\rho: G \to G'$ inducing a given homomorphism $h: LG \to LG'$ for commutative groups. We prove it now in the general case.

THEOREM 7.2.3. Let G, G' be Lie groups and $h: LG \to LG'$ a homomorphism of Lie algebras. Then there exists a local homomorphism $\rho: G \to G'$ with $L(\rho) = h$.

Note that by 6.2.11, on the domain of a canonical chart ρ must necessarily coincide with $\exp \circ h \circ \log$.

Proof: Let $k = \{(A, h(A))/A \in LG\}$. Then k is a subalgebra of $LG \times LG'$ equipped with the Lie algebra structure of definition 4.6.2. Let K be the connected Lie subgroup of $G \times G'$ with Lie algebra k. If $p: G \times G' \to G$ is the natural projection, consider the homomorphism $\lambda = p/K: K \to G$. $L(\lambda): k \to LG$ is the map given by $L(\lambda)(A, h(A)) = A$ and therefore an isomorphism. By lemma 7.2.1, λ is a local isomorphism with local inverse $\mu: G \to K$. Moreover $L(\mu)A = (A, h(A))$ for $A \in LG$. The composition of $\mu: G \to K$ with the projection $G \times G' \to G'$ gives a local homomorphism $\rho: G \to G'$. By construction $L(\rho)(A) = h(A)$ for $A \in LG$, i.e. $L(\rho) = h$, q.e.d.

Together with the unicity property of 6.2.8, the theorem expresses that L is a completely faithful functor on Lie groups and local homomorphisms to Lie algebras and Lie algebra homomorphisms.

To be able to speak strictly of unicity, we shall consider germs of local homomorphisms, i.e. we shall identify homomorphisms coinciding on a neighborhood of the identity.

We have already seen in 4.5.6 that a local isomorphism of Lie groups induces an isomorphism of Lie algebras. This is a trivial consequence of the functoriality of L. We are now able to show

THEOREM 7.2.4. Two Lie groups G and G' are locally isomorphic if and only if the Lie algebras LG and LG' are isomoprhic.

Proof: If $h: LG \to LG'$ is an isomorphism, there exists by 7.2.3 a local homomorphism $\rho: G \to G'$ inducing h, and ρ is a local isomorphism by 7.2.1, q.e.d.

This theorem is the most important fact we have proved up to now. It tells exactly which type of information onecan hope to obtain by the Lie algebra of a Lie group. Note that e.g. theorem 6.2.5 is an easy consequence of 7.2.4. To complete the study onewould like to know if every finite-dimensional Lie algebra over $\mathbb{R}$ is occuring as the Lie algebra of some Lie group. This is in fact so, but we shall not prove this here. A proof is obtained by the following theorem due to Ado: Every finite dimensional $\mathbb{R}$-Lie algebra $\mathfrak{g}$ is isomorphic to a subalgebra of the Lie algebra $\mathfrak{gl}(n, \mathbb{R})$ of $GL(n, \mathbb{R})$ for some n. The connected subgroup of $GL(n, \mathbb{R})$ corresponding to this subalgebra is a Lie group with Lie algebra isomorphic to $\mathfrak{g}$.

This shows by the way, that any Lie group is locally isomorphic to a Lie subgroup of a group $GL(n, \mathbb{R})$ for some n.

Another point to precise is the relation between local homomorphisms and (global) homomorphisms. Let $\rho: G \to G'$ be a local homomorphism. If G is connected, we know by 6.2.9 that there is at most one extension to a global homomorphism $G \to G'$. We shall make use of the following lemma on topological groups.

LEMMA 7.2.5. Let G be a connected, locally connected and simply connected topological group, G' an arbitrary topological group and $\rho: G \to G'$ a local homomorphism (of topological groups). Then there exists a unique extension of ρ to a homomorphism $\tilde{\rho}: G \to G'$.

Proof: Uniqueness is clear. To prove the existence, we define a topology on $G \times G'$. Let $V \subset G$ be a connected neighborhood of e on which ρ is defined. If $(g, g') \in G \times G'$, a fundamental system of neighborhoods is defined by $N(g, g', W) = \{(x, x')/x = wg, x' = \rho(w)g', w \in W\}$, where W is an open neighborhood of e in G with $W \subset V$. We show that the projection $p: G \times G' \to G$ is a covering of G. The map $w \rightsquigarrow (wg, \rho(w)g')$ is a homeomorphism $W \to N(g, g', W)$. If W is connected, therefore $N(g, g', W)$ is connected and $p/N(g, g', W)$ a homeomorphism: $N(g, g', W) \to W_g$. $p^{-1}(Wg)$ is the disjoint union of the $N(g, g', W)$ with $g' \in G'$. $N(g, g', W)$ are open connected subsets of this union. Therefore $G \times G'$ is locally connected and p a covering. Let $\tilde{G}$ denote the connected component of (e, e') in $G \times G'$. Then $(\tilde{G}, p/\tilde{G})$ is a covering space of G and $p/\tilde{G}$ a homeomorphism, G being simply connected. Let μ be the inverse and define $\tilde{\rho} = q \circ \mu$, where $q: G \times G' \to G'$ is the canonical projection. If $v \in V$, then $\tilde{\rho}(v) = \rho(v)$, and $\tilde{\rho}$ is an extension of ρ. It remains to show that ρ is a homomorphism. By definition of $\tilde{\rho}$, for $v \in V$ and $g \in G$ we have $\tilde{\rho}(vg) = \rho(v)\tilde{\rho}(g) = \tilde{\rho}(v)\tilde{\rho}(g)$. For $v_1 \in V$ by induction

$\tilde{\rho}((\Pi_i v_i)g) = (\Pi_i \tilde{\rho}(v_i))\tilde{\rho}(g)$ and in particular $\tilde{\rho}(\Pi_i v_i) = \Pi_i \tilde{\rho}(v_i)$. Therefore $\tilde{\rho}((\Pi_i v_i)g) = \tilde{\rho}(\Pi_i v_i)\tilde{\rho}(g)$. As V generates G $\tilde{\rho}$ is therefore a homomorphism.

Together with lemma 6.3.3 follows

PROPOSITION 7.2.6. Let G be a connected and simply connected Lie group, G' an arbitrary Lie group and $\rho: G \to G'$ a local homomorphism. Then there exists a unique extension of ρ to a homomorphism $\tilde{\rho}: G \to G'$.

Note that we have proved in proposition 5.4.8 a particular case of this proposition.

COROLLARY 7.2.7. Let G, G' be Lie groups and $h: LG \to LG'$ a homomorphism of Lie algebras. If G is connected and simply connected, then there exists a unique homomorphism $\rho: G \to G'$ with $L(\rho) = h$.

If, moreover, G' is connected and simply connected, and h an isomorphism, then ρ is an isomorphism.

Proof: To a homomorphism $h: LG \to LG'$ there exists by theorem 7.2.3 a local homomorphism $\rho: G \to G'$ inducing h . If G is connected and simply connected, ρ can, by 7.2.6, be extended uniquely to a homomorphism.

Suppose now also G' connected and simply connected. If h is an isomorphism, its inverse k is induced by a homomorphism $\lambda: G' \to G$. Now $L(\lambda \circ \rho) = 1_{LG}$ and by unicity $\lambda \circ \rho = 1_G$. Similarly

$\rho \circ \lambda = 1_{G'}$ and ρ is an isomorphism, q.e.d.

As an application, consider a commutative Lie group G. By 6.1.4 the map $\exp: LG \to G$ is a homomorphism. It is induced by the isomorphism $1_{LG}: LG \to LG$ of Lie algebras. Corollary 7.2.7 shows

PROPOSITION 7.2.8. _If G is a commutative, connected and simply connected Lie group,_ $\exp: LG \to G$ _is an isomorphism._

Remark. We mention here, without proof, the existence of a universal covering group for any connected Lie group. More precisely let G be a connected Lie group. Then there is a connected and simply connected Lie group $\widetilde{G}$ and a homomorphism and local isomorphism $\varphi: \widetilde{G} \to G$ such that $(\widetilde{G}, \varphi)$ is a covering manifold of G. $(\widetilde{G}, \varphi)$ has the following universal property. For any connected and simply connected Lie group H and homomorphism $\rho: H \to G$ there is a unique homomorphism $\widetilde{\rho}: H \to \widetilde{G}$ with $\varphi \circ \widetilde{\rho} = \rho$.

If G is a commutative connected Lie group, the pair (LG, exp) is the universal covering group of G.

Now let G be a connected Lie group, G' an arbitrary Lie group and $\rho: G \to G'$ a local homomorphism. Let $(\widetilde{G}, \varphi)$ be the universal covering group of G. Then the local homomorphism $\rho \circ \varphi: \widetilde{G} \to G'$ has by 7.2.6 a unique extension to a homomorphism $\Psi: \widetilde{G} \to G'$. If G' is connected and $(\widetilde{G}', \varphi')$ a universal covering group of G', there exists a unique homomorphism $\widetilde{\Psi}: \widetilde{G} \to \widetilde{G}'$ with $\varphi' \circ \widetilde{\Psi} = \Psi$.

Suppose in particular $\rho: G \to G'$ to be a local isomorphism of connected Lie groups G, G'. The preceding shows that $\widetilde{\Psi}: \widetilde{G} \to \widetilde{G}'$

is a local isomorphism. By corollary 7.2.7, $\mathfrak{F}$ is an isomorphism. Therefore a local isomorphism of connected Lie groups induces an isomorphism of the universal covering groups. This means that to every class of locally isomorphic connected Lie groups there corresponds a unique Lie group (up to isomorphisms), which is a universal covering group of any member of the class. Every member of the class is obtained from this universal covering group by dividing by a discrete normal subgroup (see section 7.3). By theorem 7.2.4 there is an injective map of the classes of locally isomorphic Lie groups into the classes of isomorphic $\mathbb{R}$-Lie algebras. By the above mentioned theorem of Ado this map is bijective. The problem of classifying all possible connected Lie groups is therefore decomposed in two steps. First find all $\mathbb{R}$-Lie algebras. Second find all discrete normal subgroups of a simply connected Lie group.

Consider the restricted problem of classifying all possible commutative connected Lie groups. A commutative Lie algebra is characterized by its dimension. The classification problem reduces therefore to find all discrete subgroups of a simply connected commutative Lie group. By 7.2.8, this is just the problem of finding the discrete subgroups of a finite-dimensional $\mathbb{R}$-vectorspace. We shall do this in the next section.

7.3. Discrete subgroups. Let G be a Lie group and H a subgroup. Can H be defined as a Lie subgroup of G ?

For a given topology on H (not necessarily the relative topology of H in G), such that H is a topological group, there is by 7.1.8 at most one Lie group structure inducing this topology and making H a Lie subgroup of G .

The example of the rationals $\mathbb{Q} \hookrightarrow \mathbb{R}$ shows that if we take the induced topology on H , there does not necessarily exist a Lie group structure on H inducing this topology and making H a Lie subgroup of G .

We can always consider H as a 0-dimensional manifold, making H a Lie subgroup of G . The Lie algebra of a 0-dimensional Lie group is 0 , and the subalgebra of LG corresponding to a 0-dimensional Lie subgroup therefore 0 . The example $\mathbb{Q} \hookrightarrow \mathbb{R}$ again shows that apart from this trivial manner there is, possibly, no way of turning a subgroup of a Lie group into a Lie subgroup.

DEFINITION 7.3.1. Let G be a topological group. A <u>discrete subgroup</u> H of G is a subgroup which is a discrete subspace of G. When G is a Lie group, it is natural to view a discrete subgroup H as a 0-dimensional Lie subgroup of G . Note that a discrete subgroup of a Lie group G is a closed subgroup (use the fact that G is a Hausdorff space).

Example 7.3.2. Let $0 \leq p \leq n$. Then $\mathbb{Z}^p$ is a discrete subgroup of $\mathbb{R}^n$.

We show now

PROPOSITION 7.3.3. <u>Let</u> G, G' <u>be topological groups and</u> $\rho: G \to G'$ <u>a homomorphism and local isomorphism. Then the kernel of</u>

ρ is a discrete normal subgroup of G .

Proof: There exists open neighborhoods N, N' of e, e' in G, G' such that $\rho/N: N \to N'$ is a homeomorphism. Therefore $\ker \rho \cap N = \{e\}$ and e is an isolated point of $\ker \rho$. The translations being homeomorphisms, every point of $\ker \rho$ is isolated and $\ker \rho$ is discrete. ∎

COROLLARY 7.3.4. Let G be a commutative Lie group. The kernel of the homomorphism $\exp: LG \to G$ is a discrete subgroup of the additive group of LG .

Proof: The homomorphism $\exp: LG \to G$ is by example 7.2.2 a local isomorphism.

This raises the problem of finding all discrete subgroups of the additive vectorgroup of a finite dimensional $\mathbb{R}$-vectorspace V . Every such subgroup is isomorphic to a group $\mathbb{Z}^p$, where $p \leq \dim V$. More precisely we show

LEMMA 7.3.5. Let V be a n-dimensional $\mathbb{R}$-vectorspace and D a discrete subgroup of the additive vectorgroup. Let $p \leq n$ be the dimension of the subspace generated by D . Then there exists p linearly independent vectors, $v_1, \ldots, v_p$ in V generating D .

Proof: We assume known the case $p = 1$ and prove the lemma by induction. Suppose the lemma true for all $k < p$ and let D generate a p-dimensional subspace U of V. There is a (p-1)-dimensional subspace A of U generated by elements of D. Let $v_1, \ldots, v_{p-1}$ be linearly independent vectors in V generating $D \cap A$. Now $D + A/A \cong D/D \cap A$. That this algebraic isomorphism is a topological isomorphism follows from the fact that these groups are locally compact and that D + A has a countable base (for a proof we refer to corollary 3.3 of S. Helgason [6], p. 111). Using this, we see that D + A/A is discrete. Being a subgroup of the 1-dimensional vectorspace U/A, the group D + A/A is generated by an element $v_p + A$. Then $v_1, \ldots, v_p$ are linearly independent and generate D.

We are now able to determine the structure of the commutative connected Lie groups.

THEOREM 7.3.6. Let G be a commutative connected Lie group of dimension n. Then there is an integer p, $0 \leq p \leq n$, such that $G \cong \mathbb{R}^{n-p} \times \mathbb{T}^p$.

Proof: The homomorphism $\exp: LG \to G$ is surjective by proposition 6.2.7. We have therefore an isomorphism $LG/\ker \exp \cong G$ in the algebraic sense. It is not difficult to see directly that it is an isomorphism of Lie groups. We omit this here, as we shall prove, in 7.7.6, a more general statement. Now $\ker \exp \cong \mathbb{Z}^p$ for some p with $0 \leq p \leq n$ by lemma 7.3.5. Therefore $G \cong \mathbb{R}^n/\mathbb{Z}^p$, which proves the theorem.

COROLLARY 7.3.7. Let G be a compact connected Lie group of dimension n. Then $G \cong \mathbb{T}^n$.

As mentioned at the end of section 7.2, one step in the classification problem for Lie groups consists in finding all discrete normal subgroups of a simply connected Lie group. This is greatly simplified by

PROPOSITION 7.3.8. Let H be a discrete normal subgroup of the connected topological group G. Then H is contained in the center of G.

Proof: Let $h \in H$. The map $G \to H$ defined by $g \rightsquigarrow ghg^{-1}$ is continuous. The image being connected, it must be a point and therefore equal to h, q.e.d.

7.4. Open subgroups, connectedness. Let G be a Lie group and H a subgroup which is an open subset of G. H is a submanifold and therefore a Lie subgroup of G. The Lie algebra of an open subgroup is LG, the injection being a local isomorphism. Therefore an open subgroup H of G necessarily contains G_0, as $LH = LG$ implies $H_0 = G_0$. Remember that an open subgroup is necessarily closed.

Example 7.4.1. Let V be a finite-dimensional $\mathbb{R}$-vectorspace and det: $GL(V) \to \mathbb{R}^*$ the determinant homomorphism. $\mathbb{R}^*$ being not connected, $GL(V)$ is not connected. On the other hand, any two bases of V with the same orientation (after the choice of an orientation) can be continuously transformed one into the other by automorphisms of

V. This shows readily that $\det^{-1}(\mathbb{R}^+)$ is the connected component of the identity in $GL(V)$, where $\mathbb{R}^+ = \{x \in \mathbb{R}^* / x > 0\}$.

Let H be an open subgroup of G and G/H the set of left cosets modulo H. All left cosets being open in G, the quotient topology is discrete and G/H can be considered as a 0-dimensional manifold. If in particular H is a normal subgroup of G, then G/H can be considered as a 0-dimensional Lie group.

This applies to the connected component of the identity and $\gamma = G/G_0$ is a 0-dimensional Lie group.

Example 7.4.2. Let $G = GL(V)$ be the group of automorphisms of a finite-dimensional vectorspace V. Then $\gamma = \mathbb{Z}_2$ by example 7.4.1.

As any connected component of G is diffeomorphic to G_0, the manifold G is diffeomorphic to $G_0 \times \gamma$. There is, however, no canonical diffeomorphism. Any splitting $s: \gamma \to G$ of the exact sequence $e \to G_0 \to G \to \gamma \to e$ gives rise to an isomorphism of G with the semi-direct product $G_0 x_\tau \gamma$, where $\tau: \gamma \to \mathrm{Aut}\, G_0$ is the homomorphism defined by $\tau_\epsilon = \mathfrak{J}_{s(\epsilon)}/G_0$ for $\epsilon \in \gamma$. Such a splitting exists in many particular cases.

Example 7.4.3. $G = GL(V)$ for an odd-dimensional $\mathbb{R}$-vectorspace V. The reflection at the origin of V together with the identity of V forms an isomorphic image of $\mathbb{Z}_2 = G/G_0$ in G.

In the case of a commutative group G, a splitting $s: \gamma \to G$ of the exact sequence $e \to G_0 \to G \to \gamma \to e$ defines an isomorphism of G with $G_0 x \gamma$.

Example 7.4.4. Let V be a finite dimensional $\mathbb{R}$-vectorspace. Consider a vectorsubspace U and a vector $a \in V$, $a \notin U$. Then the union of U and its translates by integer multiples of a is a Lie group G in the relative topology of V. U is the connected component of the identity of G and its group of connected components is isomorphic to $\mathbb{Z}$. The exact sequence $0 \to U \to G \to \mathbb{Z} \to 0$ has an evident splitting homomorphism $s: \mathbb{Z} \to G$ and $G \cong U \times \mathbb{Z}$.

7.5. Closed subgroups. We have seen that discrete and open subgroups of a Lie group G are Lie subgroups. Both types of subgroups are closed in G. We state now more generally

THEOREM 7.5.1. *Let* G *be a Lie group and* H *a subgroup of* G. *Suppose* H *is a closed subset of* G. *Then there exists a unique Lie group structure on* H *such that the corresponding topology is the induced topology on* H *and such that* H *is a Lie subgroup of* G.

Proof: The uniqueness statement follows from 7.1.8. Now let H be a closed subgroup of G. Let $\mathfrak{h} \subset LG$ be defined by

$$\mathfrak{h} = \{A \in LG / \exp tA \in H \text{ for every } t \in \mathbb{R}\}.$$

We first prove that $\mathfrak{h}$ is a subalgebra of $\mathfrak{h}$. $A \in \mathfrak{h}$ implies $tA \in \mathfrak{h}$ by definition of $\mathfrak{h}$. Suppose now $A, B \in \mathfrak{h}$. Then by 6.5.5, (i), $\exp t(A+B) \in H$ for every $t \in \mathbb{R}$, as H is closed. Therefore $A+B \in \mathfrak{h}$, and by 6.5.5, (ii), $\exp t^2[A,B] \in H$ for any $t \in \mathbb{R}$, showing $[A,B] \in \mathfrak{h}$. $\mathfrak{h}$ is therefore a subalgebra of LG.

Consider now the connected Lie subgroup H^* of G with $LH^* = \mathfrak{h}$. By construction of $\mathfrak{h}$ we have $\exp \mathfrak{h} \subset H$ and therefore $H^* \subset H$, H^* being generated by $\exp \mathfrak{h}$.

Let H be equipped with the relative topology of G. We shall prove that a neighborhood V of e in H^* is a neighborhood of e in H. This will prove that H^* is a topological subgroup of H (using that $H^* \hookrightarrow H$ is continuous), and taking $V = H^*$, moreover, that H^* is open in H (e being an inner point of H^* in H). Then $H^* = H_0$ as topological groups. H_0 is therefore a Lie subgroup of G. H can now be turned into a submanifold of G with the aid of translations. It is then clear that the multiplication $H \times H \to H$ will be differentiable. This is in fact sufficient to see that H is a Lie subgroup of G.

There remains to show, that a neighborhood V of e in H^* is a neighborhood of e in H. Suppose V is not a neighborhood of e in H. We show that this leads to a contradiction. There exists a sequence $c_1, \dots c_k, \dots$ in $H - V$ with $\lim_{k\to\infty} c_k = e$. Let M be a complementary subspace of $\mathfrak{h}$ in LG. By 6.3.2, there exist bounded, open, connected neighborhoods U_1, U_2 of 0 in M and $\mathfrak{h}$ respectively, such that $\phi: (A, B) \rightsquigarrow \exp A \exp B$ for $A \in M$, $B \in \mathfrak{h}$ is a diffeomorphism of $U_1 \times U_2$ onto an open neighborhood of e in G. We can, therefore, assume that $c_k = \exp A_k \exp B_k$ with $A_k \in U_1$, $B_k \in U_2$ and $\exp B_k \in V$. Then $A_k \neq 0$ and $\lim A_k = 0$.

Since $A_k \neq 0$, there exists an integer $r_k > 0$ such that $r_k A_k \in U_1$ and $(r_k + 1)A_k \notin U_1$. Now U_1 is bounded, so we can assume, passing to a subsequence, that the sequence $(r_k A_k)$ converges to a limit $A \in U_1$. Since $(r_k+1)A_k \notin U_1$ and $A_k \to 0$, A is on the boundary of U_1, in particular $A \neq 0$.

Let p, q be any integers $(q > 0)$. Then we can write $pr_k = qs_k + t_k$, where s_k, t_k are integers and $0 \leq t_k < q$. Then $\lim \frac{t_k}{q} A_k = 0$, so

$$\exp \frac{p}{q} A = \lim_k \exp \frac{p\, r_k}{q} A_k = \lim_k (\exp A_k)^{s_k} ,$$

which belongs to H. By continuity $\exp tA \in H$ for every $t \in \mathbb{R}$. But then $A \in \mathfrak{h}$, in contradiction to $A \in U_1 \subset M$ and $A \neq 0$. ∎

The previously discussed particular cases of 7.5.1, where H is either a discrete or an open subgroup of G, correspond to the case where $\mathfrak{h}$ is either 0 or equal to LG.

COROLLARY 7.5.2. Let G be a Lie group and H a closed subgroup. Let LH be the Lie algebra of H with respect to the unique Lie group structure defined in 7.5.1. Then

$$LH = \{A \in LG / \exp tA \in H \text{ for every } t \in \mathbb{R}\} .$$

Proof: The Lie algebra of H was defined in the proof of 7.5.1 by this property.

Remark. The corollary 7.5.2 is even true for an arbitrary Lie subgroup H of G which has countably many components. See S. Helgason [6], p. 108.

An important class of closed subgroups of a Lie group G are the kernels of homomorphisms starting from G.

PROPOSITION 7.5.3. *Let* $\rho: G \to G'$ *be a homomorphism of Lie groups. Then* $\ker \rho$ *is a Lie subgroup of* G *and* $L(\ker \rho) = \ker L(\rho)$, *where* $L(\rho): LG \to LG'$ *is the induced homomorphism of Lie algebras.*

Proof: $\ker \rho$ is a closed subgroup of G and therefore a Lie subgroup of G. By 7.5.2, $L(\ker \rho) = \{A \in LG / \rho(\exp tA) = e' \text{ for every } t \in \mathbb{R}\}$, e' denoting the identity of G'. By the naturality 6.1.6 of exp, $\rho(\exp tA) = e'$ for every $t \in \mathbb{R}$ is equivalent to $\exp(L(\rho)tA) = e'$ for every $t \in \mathbb{R}$. This again is equivalent to the existence of an $\epsilon > 0$, such that $L(\rho)tA = 0$ for every $|t| < \epsilon$. The latter property signifies $L(\rho)A = 0$. Therefore $L(\ker \rho) = \ker L(\rho)$, q.e.d.

This shows in particular that the kernel of the homomorphism $L(\rho): LG \to LG'$ is a Lie algebra. This is of course true for the kernel of any Lie algebra homomorphism.

We also would like the image of a homomorphism of Lie groups to be a Lie group. Indeed we have

PROPOSITION 7.5.4. *Let* $\rho: G \to G'$ *be a homomorphism of Lie groups. Suppose* G *connected. Then* $\operatorname{im} \rho$ *is a Lie subgroup of* G' *and* $L(\operatorname{im} \rho) = \operatorname{im} L(\rho)$, *where* $L(\rho): LG \to LG'$ *is the induced homomorphism of Lie algebras.*

Proof: Let H be the connected Lie subgroup of G' with $LH = \operatorname{im} L(\rho)$. H is generated by the elements $\exp(L(\rho)A)$ with $A \in LG$. Now $\rho(G)$ is generated by the elements $\rho(\exp A)$ with

$A \in LG$. But $\rho(\exp A) = \exp(L(\rho)A)$ by 6.1.6. Therefore $\rho(G) = H$, as both groups are connected.

Remark. There is the question, if the induced map $\hat{\rho}: G \to \rho(G)$ is a homomorphism, i.e. analytic. This is indeed so (see 7.7.6).

Consider now a sequence of homomorphisms of Lie groups

$$(\gamma) \quad G' \xrightarrow{\rho'} G \xrightarrow{\rho''} G''$$

and the induced sequence of homomorphisms of Lie algebras

$$(\alpha) \quad LG' \xrightarrow{L(\rho')} LG \xrightarrow{L(\rho'')} LG''$$

PROPOSITION 7.5.5. <u>Suppose</u> G' <u>connected. Then the exactness of</u> (γ) <u>implies the exactness of</u> (α).

<u>Proof:</u> If $\operatorname{im} \rho' = \ker \rho''$, then $\operatorname{im} L(\rho') = L(\operatorname{im} \rho') = L(\ker \rho'') = \ker L(\rho'')$ by 7.5.3 and 7.5.4, q.e.d.

Example 7.5.6. Let G be a connected Lie group. The exact sequence $0 \to G_e \to TG \to G \to e$ of Lie groups induces an exact sequence $0 \to G_e \to L(TG) \to LG \to 0$ of Lie algebras. Note that the natural splitting $G \hookrightarrow TG$ of the first sequence defines a splitting $LG \to L(TG)$ of the second sequence.

Observe that the converse of proposition 7.5.5 is not true, even if all groups are connected.

Example 7.5.7. Let $\rho: G \to G'$ be a homomorphism and local isomorphism. Then $0 \longrightarrow LG \xrightarrow{L(\rho)} LG' \longrightarrow 0$ is an exact sequence.

But $e \to G \to G' \to e'$ is not necessarily exact, i.e. G and G' are not necessarily isomorphic.

The following partial result is sometimes useful.

PROPOSITION 7.5.8. _Let_ $\rho: G \to G'$ _be a homomorphism of Lie groups. Suppose_ G _and_ G' _connected. Then_ ρ _is surjective if and only if_ $L(\rho): LG \to LG'$ _is surjective._

Proof: If ρ is surjective, the Lie algebra $L(\rho)LG$ of $\rho(G)$ must coincide with the Lie algebra LG' of G', which shows that $L(\rho)$ is surjective.

Suppose conversely $L(\rho)$ surjective. Then $\rho_{*g}: G_g \to G'_{\rho(g)}$ is surjective by 6.2.6 for every $g \in G$. $\rho(G)$ is therefore an open (and hence closed) subgroup of G', i.e. $\rho(G) = G'$. ▮

The condition that G' is connected cannot be omitted, as shown by the example of the inclusion of the connected component of the identity into a non-connected Lie group, which induces an isomorphism of Lie algebras.

The corresponding statement for injections is not true. We have seen in 7.1.1 that an injection $\rho: G \to G'$ induces an injection $L(\rho): LG \to LG'$. But the injectivity of $L(\rho)$ does not imply the injectivity of ρ, as shown by the example of the canonical homomorphism $\mathbb{R} \to \mathbb{T}$.

7.6. Closed subgroups of the full linear group. Let V be a finite dimensional $\mathbb{R}$-vectorspace and $GL(V)$ the group of linear automorphisms. We shall consider some closed subgroups of $GL(V)$.

Let $\phi: V \times V \to \mathbb{R}$ be a bilinear and non-degenerated form on V. Let H be the subgroup of $GL(V)$ leaving ϕ invariant:

$$H = \{g \in GL(V) / \phi(gv, gw) = \phi(v, w) \quad \text{for any} \quad v, w \in V\} \quad .$$

Consider for fixed $v, w \in V$ the map $GL(V) \to GL(V) \times GL(V) \to V \times V \to \mathbb{R}$ defined by $g \rightsquigarrow (g, g) \rightsquigarrow (gv, gw) \rightsquigarrow \phi(gv, gw)$. As ϕ is continuous, this map is continuous. Therefore the set $S(v, w) = \{g \in GL(V) / \phi(gv, gw) = \phi(v, w)\}$ is closed in $GL(V)$. Now

$$H = \bigcap_{v, w \in V} S(v, w)$$

This shows that H is a closed subgroup of $GL(V)$.

We identify the Lie algebra of $GL(V)$ with $\mathcal{L}(V)$ (see proposition 4.3.8). Then we have the following characterization of LH.

PROPOSITION 7.6.1.

$$LH = \{A \in \mathcal{L}(V) / \phi(Av, w) + \phi(v, Aw) = 0 \quad \text{for} \quad v, w \in V\}$$

Proof: Let $A \in LH$. Then $\exp tA \in H$ for any $t \in \mathbb{R}$ and $\phi(\exp(tA) v, \exp(tA) w) = \phi(v, w)$ for $v, w \in V$. Differentiating with respect to t we obtain for $t = 0$, $\phi(Av, w) + \phi(v, Aw) = 0$.

Suppose conversely that $A \in \mathcal{L}(V)$ satisfies this condition. Denote by A^* the adjoint linear map of A with respect to ϕ, characterized by $\phi(Av, w) - \phi(v, A^* w) = 0$. The hypothesis can therefore be expressed by $A^* = -A$. We shall show that $(\exp tA)^* = (\exp tA)^{-1}$ for every $t \in \mathbb{R}$, which means $\exp tA \in H$ for every $t \in \mathbb{R}$. This implies

$\Lambda \in LH$.

There remains to show that $A^* = -A$ implies $(\exp tA)^* = (\exp tA)^{-1}$. But this follows from the expression $\exp tA = \Sigma_{h=0}^{\infty} \frac{(tA)^n}{n!}$ given in 6.1.5.

Exercise. Deduce also proposition 7.6.1 from 6.4.5.

If ϕ is moreover symmetric,the group H is the orthogonal group of V with respect to ϕ and denotea $O(V,\phi)$. The Lie algebra consists of the operators of V which are antiselfadjoint with respect to ϕ .

Example 7.6.2. Consider a 3-dimensional $\mathbb{R}$-vectorspace V with a Euclidean metric ϕ . Let e_1, e_2, e_3 be an orthonormal base with a positive orientation. Then V can be turned into a Lie algebra by the definition $[e_1, e_2] = e_3$, $[e_1, e_3] = -e_2$, $[e_2, e_3] = e_1$. We identify $\mathfrak{L}(V)$ with the Lie algebra of $GL(V)$ (see 4.3.8). Let $a \in V$ and consider $A \in \mathfrak{L}(V)$ defined by $Av = [a, v]$ for $v \in V$. Then $\phi(Av, w) + \phi(v, Aw) = \phi([a, v], w) + \phi(v, [a, w]) = 0$, as is seen by the interpretation of $([a, v], w)$ as the oriented volume of the parallelepiped defined by a, v, w . Therefore A is contained in the Lie algebra $LO(v, \phi)$ of the orthogonal group $O(V, \phi)$ with respect to ϕ . The map $\mathfrak{J}: V \to LO(V, \phi)$ defined by $a \rightsquigarrow A$ is linear. But $[a, v] = 0$ for every $v \in V$ implies $a = 0$ for this particular Lie algebra structure defined on V . This means that $\mathfrak{J}$ is injective. But both V and $LO(V, \phi)$ have dimension 3 , so $\mathfrak{J}$ is a linear isomorphism. We finally verify that $\mathfrak{J}: V \to LO(V, \beta)$ is an isomorphism of Lie algebras. Let

$A_i v = [a_i, v]$ for $v \in V$, $i = 1, 2$. Then

$[A_1, A_2]v = A_1A_2v - A_2A_1v = [a_1, [a_2, v]] - [a_2, [a_1, v]] = [[a_1, a_2], v]$,

using the Jacobi identity.

We have therefore seen that V is isomorphic to the Lie algebra of the group $O(V, \phi)$ by the map $\mathfrak{J}: V \to LO(V, \phi)$, defined by $a \rightsquigarrow A$, where $Av = [a, v]$ for every $v \in V$.

To see the geometric significance of the correspondence $a \rightsquigarrow A$, consider the 1-parameter subgroup α of $O(V, \phi)$ defined by A, satisfying $\dot{\alpha}_t = A\alpha_t$ by 5.4.5. Let $v \in V$ and $v_t = \alpha_t v$. Then $\dot{v}_t = \dot{\alpha}_t v = A\alpha_t v = Av_t$ can be written as $\dot{v}_t = [a, v_t]$. This shows that the 1-parameter subgroup α_t of $O(V, \phi)$ defined by A is the 1-parameter group of rotations of V with rotation axis a.

$O(V, \phi)$ operates in V by automorphisms of the Lie algebra structure defined in V. The isomorphisms $\mathfrak{J}: V \to LO(V, \phi)$ defines therefore a representation of $O(V, \phi)$ in $LO(V, \phi)$. If $\sigma: O(V, \phi) \to \operatorname{Aut} LO(V, \phi)$ denotes this representation, then σ is defined by $(\sigma_g A)(v) = [ga, v]$ for $g \in O(V, \phi)$, $a \in V$, $A = \mathfrak{J}(a)$ and every $v \in V$. This representation σ is just the adjoint representation of $O(V, \phi)$, because

$$(\sigma_g A)(v) = [ga, v] = g[a, g^{-1}v]$$
$$= g(A(g^{-1}v)) = (gAg^{-1})(v) \qquad \text{for every } v \in V$$

and therefore $\sigma_g A = gAg^{-1}$.

Let V be again of arbitrary finite dimension and ϕ a non-degenerated bilinear and symmetric form on V. Suppose ϕ positive definite. By the same argument as for $GL(V)$ (see example 7.4.1), one shows that the connected component of the identity is the kernel of the homomorphism

$\det : O(V, \phi) \to \mathbb{R}^*$. This group is denoted by $SO(V, \phi)$.

PROPOSITION 7.6.3. Let V be a finite dimensional $\mathbb{R}$ vector-space, ϕ a positive definite, symmetric bilinear form on V , $O(V, \phi)$ the orthogonal group of V with respect to ϕ and $SO(V, \phi)$ the group of orthogonal operators with determinant 1 . Then $O(V, \phi)$ and $SO(V, \phi)$ are compact.

Proof: $SO(V, \phi)$ is an open subgroup of $O(V, \phi)$ and therefore closed. Hence it is sufficient to prove $O(V, \phi)$ compact. Now $O(V, \phi)$ is a closed subgroup of $GL(V)$. $GL(V)$ being an open subset of $\mathcal{L}(V)$, $O(V, \phi)$ is also closed in $\mathcal{L}(V)$. It suffices therefore to show that $O(V, \phi)$ is bounded in $\mathcal{L}(V)$.

Now let $| \ | : \mathcal{L}(V) \to \mathbb{R}$ be the norm on $\mathcal{L}(V)$ defined with respect to ϕ by

$$|A| = \sup_{v \neq 0, v \in V} \frac{\phi(Av, Av)^{1/2}}{\phi(v, v)^{1/2}} .$$

Then any $g \in O(V, \phi)$ satisfies $|g| = 1$ and $O(V, \phi)$ is bounded in $\mathcal{L}(V)$.

Let now $\phi : V \times V \to \mathbb{R}$ be a skew-symmetric bilinear and non-degenerated form on V (V of even dimension). The subgroup of $GL(V)$ leaving ϕ invariant is the symplectic group of V with respect to ϕ, denoted $Sp(V, \phi)$. As there is essentially a unique ϕ of that type, any two symplectic groups of V are isomorphic. The Lie algebra of $Sp(V, \phi)$ consists, according to 7.6.1, of the operators of V which are antiselfadjoint with respect to ϕ .

Consider now the homomorphism $\det: GL(V) \to \mathbb{R}^*$. The kernel is denoted by $SL(V)$.

PROPOSITION 7.6.4. Let V be a finite-dimensional $\mathbb{R}$-vector-space. The set of operators with trace 0 is a Lie algebra. It is the Lie algebra of $SL(V)$.

Proof: By 4.5.11, $L(\det) = \mathrm{tr}$. Now 7.5.3 shows $L(\ker \det) = L(SL(V)) = \ker \mathrm{tr}$, q.e.d.

7.7. Coset spaces, factor groups.

THEOREM 7.7.1. Let G be a Lie group and H a closed subgroup. Let G/H be the orbitspace of the operation of H on G by right-translations(see 2.2.3). Consider the natural operation of G on G/H (section 1.4). Then there exists a unique structure of analytic manifold on G/H , inducing the quotient topology and making it a G-manifold.

Let H be equipped with the structure of Lie group of 7.5.1. Let M be a vector-subspace of LG such that $LG = M \oplus LH$. Denote by $p: G \to G/H$ the canonical projection. Then theorem 7.7.1 is based on the following lemma, which proof we omit (see S. Helgason, [6], p. 113) .

LEMMA 7.7.2. There exists a neighborhood U of 0 in M , such that $\exp/U: U \to \exp(U)$ is a homeomorphism and $p/\exp(U): \exp(U) \to p(\exp U)$ is a homeomorphism onto a neighborhood of $p(e)$ in G/H .

The structure of analytic manifold on G/H is then defined as follows. If N_0 denotes the interior of $p(\exp U)$ and $\mathring{U}$ the interior of U, then $(\exp/\mathring{U})^{-1} \circ (p/\exp(\mathring{U}))^{-1} : N_0 \to \mathring{U} \subset M$ is a chart at $p(e) \in G/H$. Now G operates by homeomorphisms on G/H, so that this defines also charts at any point of G/H. It is to show that these charts are compatible, i.e. define an analytic structure on G/H. Then, by construction, G operates by analytic maps on G/H. The unicity of the analytic structure on G/H as announced in 7.7.1 follows from the last statement in

PROPOSITION 7.7.3. *Let* X *be a G-manifold with respect to a transitive operation* $\tau: G \to \mathrm{Aut}\, X$. *Select* $x_0 \in X$ *and let* H *be the isotropy group of* x_0. *Consider the map* $\varphi: G/H \to X$ *defined by* $\varphi(gH) = \tau_g(x_0)$. *Let* G/H *have the analytic manifold structure defined above. Then* φ *is differentiable. If* φ *is a homeomorphism, then it is a diffeomorphism.*

Proof: We use N_0 and $\mathring{U}$ with the same meaning as before and write $B = \exp(\mathring{U})$. Then the homeomorphism $p/B: B \to N_0$ permits defining B as a submanifold of G, making the injection $\iota: B \hookrightarrow G$ differentiable. Denote by $\Psi: G \to X$ the map $\Psi(g) = \tau_g(x_0)$. Then $\varphi/N_0 = \Psi \circ \iota \circ (p/B)^{-1}$ and φ is therefore differentiable.

Now suppose φ to be a homeomorphism (see remark below). φ will be a diffeomorphism if the tangent linear map of φ at any point is an isomorphism. φ being an equivariance (see 1.4.10), it is sufficient to prove this for the point x_0. Now the decomposition $\varphi/N_0 = \Psi \circ \iota \circ (p/B)^{-1}$ shows that it is sufficient to prove $\Psi_{*e}: G_e \to T_{x_0}(X)$ to be surjective. We shall prove $\ker \Psi_{*e} = H_e$. Then rank $\Psi_{*e} = \dim G - \dim \ker \Psi_{*e}$ $= \dim G - \dim H = \dim G/H = \dim X$ (the last equality because a is a homeomorphism), and this will finish the proof.

There remains to show that $\ker \Psi_{*e} = H_e$. H being the isotropy group of x_0, clearly $H_e \subset \ker \Psi_{*e}$. Let conversely $A_e \in \ker \Psi_{*e}$. Consider the Killingvectorfield A^* on X defined by A_e, respectively the corresponding $A \in RG$, i.e. $A^* = \sigma(a)$ in the notation of theorem 5.6.2. Then A and A^* are Ψ-related (see the proof of 5.6.2), which shows $\Psi_{*e} A_e = A^*_{x_0} = 0$. Then by 6.3.1, $\exp tA_e \in H$ for every $t \in \mathbb{R}$. Thus $A_e \in H_e$ in view of 7.5.3, q.e.d.

Remark. The map $\varphi: G/H \to X$ defined in 7.7.3 is in fact a homeomorphism, if G has countably many components. Under this condition, the argument in the proof above shows, for an arbitrary G-manifold X (with a not necessarily transitive operation) and $x_0 \in X$, that $\varphi: G/H \to X$ is a diffeomorphism onto the orbit of x_0.

Before turning to the case where H is a normal subgroup of G, we give a definition.

DEFINITION 7.7.4. Let $\mathfrak{g}$ be a Lie algebra over a ring Λ. An <u>ideal</u> $\mathfrak{h}$ of $\mathfrak{g}$ is a vectorsubspace of $\mathfrak{g}$ satisfying $[A, B] \in \mathfrak{h}$, for every $A \in \mathfrak{g}$, $B \in \mathfrak{h}$.

If $\mathfrak{h}$ is an ideal of $\mathfrak{g}$, the quotient vectorspace $\mathfrak{g}/\mathfrak{h}$ is canonically equipped with a Lie algebra structure, and $\mathfrak{h}$ is the kernel of the canonical homomorphism $\mathfrak{g} \to \mathfrak{g}/\mathfrak{h}$. Conversely, the kernel $\mathfrak{h}$ of a Lie algebra homomorphism, with domain $\mathfrak{g}$, is an ideal of $\mathfrak{g}$, and the vectorspace isomorphism of $\mathfrak{g}/\mathfrak{h}$ with the image is a Lie algebra isomorphism.

We prove now

PROPOSITION 7.7.5. Let H be a closed normal subgroup of the Lie group G . The factorgroup G/H with the manifold structure defined in 7.7.1 is a Lie group. The canonical homomorphism $p: G \to G/H$ induces $L(p): LG \to L(G/H)$ with kernel LH , such that $LG/LH \cong L(G/H)$.

Proof: The factorgroup G/H is a topological group with respect to the quotient topology. Consider the unique manifold structure of 7.7.1 on G/H such that the map $G \times G/H \to G/H$ given by $(g, xH) \rightsquigarrow gxH$ is analytic. There remains to show that the group operations in G/H are analytic, which is immediate. By construction of the manifold structure on G/H , $p: G \to G/H$ is analytic , and therefore a homomorphism of Lie groups.

Consider $L(p): LG \to L(G/H)$. By 7.5.3 ker $L(p) = L(\ker p) = LH$. Therefore $L(p)$ induces an isomorphism $LG/LH \cong L(G/H)$.

Note that if H is a normal subgroup of G which is not closed, the factorgroup is not Hausdorff.

As a consequence we obtain

PROPOSITION 7.7.6. <u>Let</u> $\rho: G \to G'$ <u>be a homomorphism of Lie groups.</u> <u>Suppose</u> G <u>connected.</u> <u>Consider the canonical map</u> $\tilde{\rho}: G/\ker\rho \to \rho(G)$ <u>induced by</u> ρ . <u>Then</u> $\tilde{\rho}$ <u>is an isomorphism of Lie groups, where</u> $G/\ker\rho$ <u>is equipped with the Lie group structure of</u> 7.7.5 <u>and</u> $\rho(G)$ <u>with that of</u> 7.5.4. <u>This shows in particular that the map</u> $\hat{\rho}: G \to \rho(G)$ <u>induced by</u> ρ <u>is analytic.</u>

<u>Proof:</u> Consider the commutative diagram

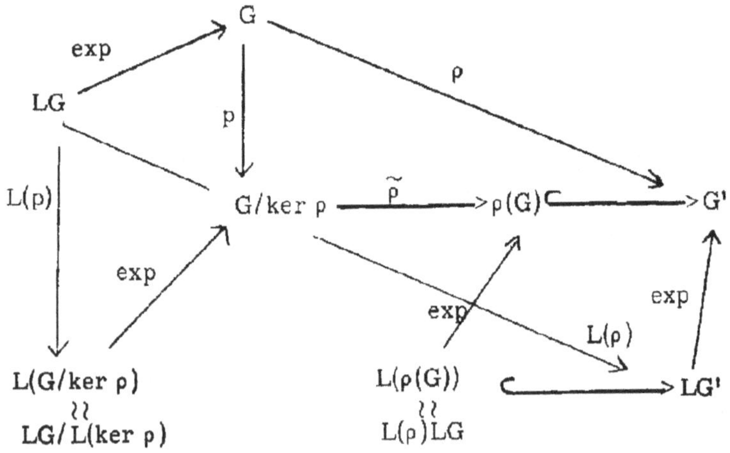

There is at most one map $\gamma: L(G/\ker\rho) \to L(\rho(G))$ filling in, as $L(p)$ is surjective and $L(\rho(G)) \to LG'$ is injective. Consider the canonical isomorphism $\gamma: LG/L(\ker\rho) \to L(\rho)LG$ induced by $L(\rho): LG \to LG'$. γ makes the diagram commutative. This proves that $\tilde{\rho}$ is analytic at e , being just γ in canonical charts. Hence $\tilde{\rho}$ is everywhere analytic Moreover $L(\tilde{\rho}) = \gamma$ is an isomorphism and therefore $\tilde{\rho}$ an isomorphism of Lie groups. The map $\hat{\rho}: G \to \rho(G)$ is the composition $\tilde{\rho} \circ p$ of analytic homomorphisms and hence analytic.

CHAPTER 8. GROUPS OF AUTOMORPHISMS

8.1. The automorphism group of an algebra. Let $\mathfrak{A}$ be a finite dimensional $\mathbb{R}$-algebra, i.e. a vectorspace with a bilinear map $\mathfrak{A} \times \mathfrak{A} \to \mathfrak{A}$. $GL(\mathfrak{A})$ is the group of automorphisms of the underlying vectorspace. Aut $\mathfrak{A}$ is the group of automorphisms of the algebra $\mathfrak{A}$. Then Aut $\mathfrak{A} \subset GL(\mathfrak{A})$.

Example 8.1.1. $\mathfrak{A}$ a $\mathbb{R}$-Lie algebra.

LEMMA 8.1.2. Aut $\mathfrak{A}$ is a closed subgroup of $GL(\mathfrak{A})$.

Proof: Let $A, B \in \mathfrak{A}$ and consider the map

$$GL(\mathfrak{A}) \to GL(\mathfrak{A}) \times GL(\mathfrak{A}) \to \mathfrak{A} \times \mathfrak{A} \to \mathfrak{A}$$

defined by $\varphi \rightsquigarrow (\varphi, \varphi) \rightsquigarrow (\varphi A, \varphi B) \rightsquigarrow \varphi A . \varphi B$. The multiplication $\mathfrak{A} \times \mathfrak{A} \to \mathfrak{A}$ being continuous ($\mathfrak{A}$ is finite dimensional), this map is continuous. The set $S(A,B) = \{\varphi \in GL(\mathfrak{A}) / \varphi A . \varphi B = \varphi(A.B)\}$ is the inverse image of $\varphi(A.B)$ under this map and therefore closed in $GL(\mathfrak{A})$. Now $\mathrm{Aut}(\mathfrak{A}) = \bigcap_{A,B \in \mathfrak{A}} S(A,B)$ and therefore Aut $\mathfrak{A}$ closed in $GL(\mathfrak{A})$.

By 7.5.2 we have therefore

PROPOSITION 8.1.3. Let $\mathfrak{A}$ be a finite dimensional $\mathbb{R}$-algebra. Then Aut $\mathfrak{A}$ is a closed Lie subgroup of $GL(\mathfrak{A})$. Its Lie algebra $\partial(\mathfrak{A})$ is characterized by

$$\partial(\mathfrak{A}) = \{D \in \mathfrak{L}(\mathfrak{A}) / \exp tD \in \mathrm{Aut}\, \mathfrak{A} \quad \text{for every} \quad t \in \mathbb{R}\} .$$

Here $\mathfrak{L}(\mathfrak{A})$ denotes the Lie algebra of endomorphisms of the underlying vectorspace of $\mathfrak{A}$.

DEFINITION 8.1.4. A derivation D of $\mathfrak{A}$ is an element $D \in \mathfrak{L}(\mathfrak{A})$ satisfying

$$D(A.B) = DA.B + A.DB \quad \text{for every} \quad A, B \in \mathfrak{A}.$$

PROPOSITION 8.1.5. The Lie algebra $\partial(\mathfrak{A})$ is the set of derivations of $\mathfrak{A}$.

Proof: Let $D \in \partial(\mathfrak{A})$. By 8.1.3

$$\exp tD(A.B) = (\exp tD.A).(\exp tD.B) \quad \text{for every } A, B \in \mathfrak{A}, \; t \in \mathbb{R}.$$

Differentiating with respect to t we obtain for $t = 0$

$$D(A.B) = DA.B + A.DB \quad \text{and} \quad D \text{ is a derivation of } \mathfrak{A}.$$

Conversely, let D be a derivation of $\mathfrak{A}$. By induction we get

$$D^n(A.B) = \sum_{i+j=n} \frac{n!}{i!\,j!} D^iA.D^jB \qquad i \geqq 0, \; j \geqq 0$$

(For $n = 0$ this is true, D^0 being the identity.)

Now, by 6.1.5 we have

$$\exp tD = \sum_{n=0}^{\infty} \frac{(tD)^n}{n!}.$$

Therefore

$$\exp tD(A.B) = \sum_{n=0}^{\infty} \frac{(tD)^n}{n!}(A.B)$$

$$= \left(\sum_{i=0}^{\infty} \frac{(tD)^iA}{i!}\right)\left(\sum_{j=0}^{\infty} \frac{(tD)^jB}{j!}\right) = (\exp tD.A).(\exp tD.B)$$

and $\exp tD \in \operatorname{Aut}\mathfrak{A}$ for every $t \in \mathbb{R}$. By 8.1.3 this shows $D \in \partial(\mathfrak{A})$.

Remark. The fact that the set of derivations of $\mathfrak{A}$ is a subalgebra of the Lie algebra $\mathfrak{L}(\mathfrak{A})$ follows also directly and is true without any restriction on the dimension of $\mathfrak{A}$. Proposition 8.1.5 suggests, in this case also, viewing heuristically the Lie algebra of derivations as the Lie algebra of the group of automorphisms of $\mathfrak{A}$.

In particular, let X be a manifold and CX the $\mathbb{R}$-algebra of functions $X \to \mathbb{R}$. The Lie algebra DX of vectorfields on X is the Lie algebra of derivations of CX. Now by 4.1.3, Aut X can be identified with Aut CX. So DX can be thought of as the Lie algebra of Aut X, as we have indicated at several places before.

Let now X be a G-manifold with respect to an operation $\tau: G \to \operatorname{Aut} X$. It induces an operation $\tau^*: G \to \operatorname{Aut} CX$. Consider the homomorphism $\sigma: RG \to DX$ of 5.6.2. It can be thought of being the homomorphism of Lie algebras induced by the homomorphism τ^*.

8.2. <u>The adjoint representation of a Lie algebra.</u> We begin with some remarks for a Lie algebra $\mathfrak{g}$ over a ring Λ.

Any element $A \in \mathfrak{g}$ gives rise to a linear map $\operatorname{ad} A: \mathfrak{g} \to \mathfrak{g}$ by the definition $(\operatorname{ad} A)(B) = [A, B]$.

LEMMA 8.2.1. ad A *is a derivation of* $\mathfrak{g}$.

Proof: The Jacobi identity can be written in the form

$$[A,[B_1,B_2]] = [[A,B_1],B_2] + [B_1,[A,B_2]]$$

which proves the desired result.

DEFINITION 8.2.2. Let $\mathfrak{g}$ be a Lie algebra. The inner derivation of $\mathfrak{g}$ defined by $A \in \mathfrak{g}$ is the map $\operatorname{ad} A: \mathfrak{g} \to \mathfrak{g}$.

Consider the map $\operatorname{ad}: \mathfrak{g} \to \mathcal{L}(\mathfrak{g})$ into the Lie algebra of endomorphisms of $\mathfrak{g}$.

LEMMA 8.2.3. $\operatorname{ad}: \mathfrak{g} \to \mathcal{L}(\mathfrak{g})$ is a homomorphism of Lie algebras.

Proof: This is again a consequence of the Jacobian identity, namely

$$\begin{aligned}
(\operatorname{ad}[A_1,A_2])(B) &= [[A_1,A_2],B] \\
&= [A_1,[A_2,B]] - [A_2,[A_1,B]] \\
&= (\operatorname{ad} A_1 \circ \operatorname{ad} A_2)(B) - (\operatorname{ad} A_2 \circ \operatorname{ad} A_1)(B) \\
&= [\operatorname{ad} A_1, \operatorname{ad} A_2](B)\,, \qquad \text{q.e.d.}
\end{aligned}$$

We have seen before that $\operatorname{ad}(\mathfrak{g}) \subset \partial(\mathfrak{g})$, where $\partial(\mathfrak{g})$ is the Lie algebra of derivations of $\mathfrak{g}$, subalgebra of the Lie algebra $\mathcal{L}(\mathfrak{g})$. We shall also write $\operatorname{ad}: \mathfrak{g} \to \partial(\mathfrak{g})$ for the homomorphism induced by $\operatorname{ad}: \mathfrak{g} \to \mathcal{L}(\mathfrak{g})$.

DEFINITION 8.2.4. Let $\mathfrak{g}$ be a Lie algebra. The homomorphism $\operatorname{ad}: \mathfrak{g} \to \partial(\mathfrak{g})$ is called the adjoint representation of $\mathfrak{g}$. It is a representation of g in $\mathfrak{g}$.

The image of this homomorphism is the set of inner derivations of $\mathfrak{g}$, which therefore forms a Lie algebra.

Let $\mathfrak{z}(\mathfrak{g})$ denote the kernel of this homomorphism. It is an ideal of $\mathfrak{g}$, called the center of $\mathfrak{g}$, and is characterized by $A \in \mathfrak{z}(\mathfrak{g})$ if and only if $[A,B] = 0$ for any $B \in \mathfrak{g}$.

Now let G be a Lie group. By 8.1.3 Aut LG is a Lie group. The Lie algebra is by 8.1.4 the set of derivations $\partial(LG)$ of LG.

Consider the adjoint representation of G in LG, $Ad: G \to Aut\ LG$. By the preceding it induces a homomorphism $L(Ad): LG \to \partial(LG)$.

THEOREM 8.2.5. $L(Ad) = ad$.

Proof: Let $A \in LG$. Then

$$L(Ad)A = \frac{d}{dt}\{Ad\ exp\ tA\}_{t=0}$$

by definition of $L(Ad)$. But by 5.5.8, the second member is just ad A, q.e.d.

COROLLARY 8.2.6. Let G be a connected Lie group. Then $L(Ad\ G) = ad\ (LG)$.

Proof: By 7.5.4 we have $L(Ad\ G) = L(\text{im}\ Ad) = \text{im}\ L(Ad) = \text{im}\ ad = ad\ LG$, q.e.d.

COROLLARY 8.2.7. Let G be a connected Lie group. Then the center ZG is a Lie subgroup of G. Its Lie algebra is the center of LG.

Proof: By 6.2.10 we know that $ZG = \ker Ad$. By 7.5.3, ZG is therefore a closed Lie subgroup of G with Lie algebra $L(ZG) = L(\ker Ad) = \ker L(Ad) = \ker ad$. But $\ker ad$ is the center of LG.

Note that $Ad: G \to Aut\, LG$ induces an isomorphism $G/ZG \cong Ad\, G$ of Lie groups (see 7.7.6).

COROLLARY 8.2.8. $\exp ad\, A = Ad \exp A$ for $A \in LG$.

Proof: This is the naturality of exp.

We shall make use of the following two lemmas.

LEMMA 8.2.9. Let V be a finite dimensional $\mathbb{R}$-vectorspace, G a connected Lie group, $\tau: G \to GL(V)$ a representation of G in V and $L(\tau): LG \to \mathcal{L}(V)$ the induced representation of LG in V. A vectorspace $W \subset V$ is G-invariant if and only if it is LG-invariant. Precisely: $\tau_g W \subset W$ for every $g \in G$ if and only if $(L(\tau)A)W \subset W$ for every $A \in LG$.

Proof: Suppose $W \subset V$ to be G-invariant and let $A \in LG$, $w \in W$.

$$(L(\tau)A)w = \frac{d}{dt}\left\{\tau_{\exp tA}\right\}\Big|_{t=0} w = \frac{d}{dt}\left\{\tau_{\exp tA}\, w\right\}\Big|_{t=0},$$

which is the tangent vector of the curve $t \rightsquigarrow \tau_{\exp tA} w$ in W for $t = 0$, and therefore $(L(\tau)A)w \in W$.

Suppose conversely $W \subset V$ to be LG-invariant, i.e. for every $A \in LG$ we have $(L(\tau)A)W \subset W$. Now by 6.1.5 follows immediately

$(\exp L(\tau)A)W \subset W$. By the naturality6.1.6 of exp this is equivalent to $\tau_{\exp A} W \subset W$ for every $A \in LG$. Let $\tilde{G} = \{g \in G / \tau_g W \subset W\}$. Then $\tilde{G}$ is a subgroup of G . By the preceding, $\tilde{G}$ contains a neighborhood of e in G and therefore $\tilde{G} = G$.

LEMMA 8.2.10. Let G, G' be Lie groups and H, H' connected Lie subgroups of G, G' respectively. Let $\rho: G \to G'$ be a homomorphism. Then $\rho(H) \subset H'$ if and only if $L(\rho)LH \subset LH'$.

Proof: Clear from 7.5.4.

Let $\mathfrak{g}$ be a Lie algebra. The definition 7.7.4 of an ideal of $\mathfrak{g}$ can be restated by saying that a vectorspace $\mathfrak{h} \subset \mathfrak{g}$ is an ideal if and only if $\mathfrak{h}$ is $\mathrm{ad}\,\mathfrak{g}$-invariant.

Let G be a Lie group. We have seen in 7.7.5 that the Lie algebra of a closed normal subgroup of a Lie group G is an ideal of LG . We are now able to prove

PROPOSITION 8.2.11. Let G be a connected Lie group and H a connected Lie subgroup of G . Then H is a normal subgroup of G if and only if LH is an ideal of LG .

Proof: LH is an ideal of LG if and only if LH is ad LG-invariant. In view of $L(\mathrm{Ad}) = \mathrm{ad}$ and 8.2.9, this is equivalent to the Ad G-invariance of LH, i.e. $\mathrm{Ad}\, g\, LH \subset LH$ for every $g \in G$. But $\mathrm{Ad}\, g = L(\mathfrak{I}_g)$ by definition, and using 8.2.10 we see that $\mathrm{Ad}\, g\, LH \subset LH$ if and only if $\mathfrak{I}_g(H) \subset H$. Therefore LH is an ideal of LG if and only if $\mathfrak{J}_g(H) \subset H$ for every $g \in G$, q.e.d.

COROLLARY 8.2.12. Let G be a connected Lie group. Then Ad G is a normal Lie subgroup of $(\mathrm{Aut}\ LG)_0$.

Proof: Ad G is connected and therefore contained in $(\mathrm{Aut}\ LG)_0$. Now $L(\mathrm{Ad}) = \mathrm{ad}\ LG$ by 8.2.6. In view of 8.2.11 there is only to show that ad LG is an ideal of $L(\mathrm{Aut}\ LG) = \partial(LG)$. This is true for an arbitrary Lie algebra $\mathfrak{g}$. Let namely $D \in \partial(\mathfrak{g})$, $A \in \mathfrak{g}$. Then there is to show $[D, \mathrm{ad}\ A] \in \mathrm{ad}\ \mathfrak{g}$. For $B \in \mathfrak{g}$ we have $[D, \mathrm{ad}\ A]B = D[A, B] - [A, DB] = [DA, B] = (\mathrm{ad}\ DA)B$, which shows $[D, \mathrm{ad}\ A] = \mathrm{ad}\ DA$.

Remark. The group Ad G is not necessarily closed in Aut LG.

8.3. The automorphism group of a Lie group. Let G be a Lie group and Aut G the group of automorphisms (of the Lie group structure; however, remember 6.3.4). The functor L defines a homomorphism $L: \mathrm{Aut}\ G \to \mathrm{Aut}\ LG$ into the group of automorphisms of LG. If G is connected, 6.2.9 shows that this homomorphism is injective.

Example 8.3.1. Consider the Lie group $\mathbb{T} = \mathbb{R}/\mathbb{Z}$. Then $\mathrm{Aut}\ \mathbb{T} \to \mathrm{Aut}\ (L\mathbb{T}) = GL(\mathbb{R}) = \mathbb{R}^*$ is injective. In fact, $\mathrm{Aut}\ \mathbb{T} = \{1_{\mathbb{T}}, -1_{\mathbb{T}}\}$, where $-1_{\mathbb{T}}$ denotes the map induced on $\mathbb{T}$ by $-1_{\mathbb{R}}$ (see 8.3.4).

If G is connected and simply connected, the homomorphism $L: \mathrm{Aut}\ G \to \mathrm{Aut}\ LG$ is an isomorphism by 7.2.7.

Example 8.3.2. $G = \mathbb{R}$. Then $\operatorname{Aut} \mathbb{R} = \mathbb{R}^*$.

More generally, let G be a commutative connected and simply connected Lie group. Then $\exp: LG \to G$ is an isomorphism by 7.2.8. $\operatorname{Aut} G \to \operatorname{Aut} LG = GL(LG)$ is an isomorphism.

Let G be a commutative and connected Lie group. An automorphism φ of G defines an automorphism $L(\varphi)$ of LG . Consider the homomorphism $\exp: LG \to G$. Then $L(\varphi) \ker \exp \subset \ker \exp$ in view of the commutative diagram

$$\begin{array}{ccc} LG & \xrightarrow{L(\varphi)} & LG \\ {\scriptstyle \exp}\downarrow & & \downarrow{\scriptstyle \exp} \\ G & \xrightarrow{\varphi} & G \end{array}$$

We have proved half of

PROPOSITION 8.3.3. *Let* G *be a commutative connected Lie group. Then the image of the homomorphism* $L: \operatorname{Aut} G \to GL(LG)$ *consists of the automorphism* $\tilde{\varphi}$ *of* LG *with* $\tilde{\varphi} \ker \exp \subset \ker \exp$.

Proof: We have to show that given $\tilde{\varphi} \in GL(LG)$ with $\tilde{\varphi} \ker \exp \subset \ker \exp$, there exists $\varphi \in \operatorname{Aut} G$ with $L(\varphi) = \tilde{\varphi}$. But $(\exp \circ \tilde{\varphi}) \ker \exp = e$ implies that there exists a factorization $\varphi: G \to G$ of $\exp \circ \tilde{\varphi}$ through $\exp$ and clearly $L(\varphi) = \tilde{\varphi}$.

Remark. There is a similar characterization of Aut G for an arbitrary connected Lie group. One has only to consider the universal covering group $\tilde{G}$ and the covering homomorphism $\tilde{G} \to G$.

Proposition 8.3.3 allows us to determine Aut G, as $G \cong LG/\ker \exp$. We show

PROPOSITION 8.3.4. Let $G = \mathbb{T}^n$. Then $\operatorname{Aut} G \cong \operatorname{Aut} \mathbb{Z}^n$.

Proof: We denote by $\mathbb{Z}^n$ a subgroup of LG isomorphic to $\mathbb{Z}^n$. Then $\mathbb{T}^n \cong LG/\mathbb{Z}^n$. Now $\operatorname{Aut} \mathbb{T}^n = \{\Psi \in GL(LG) / \Psi(\mathbb{Z}^n) \subset \mathbb{Z}^n\}$. This shows $\operatorname{Aut} \mathbb{T}^n \cong \operatorname{Aut} \mathbb{Z}^n$, q.e.d.

Appendix. Categories and functors

Definition. A category $\mathfrak{K}$ consists of

(i) a class of objects A, B, C, ... ;

(ii) for each pair (A, B) of objects a set [A, B], which elements are called morphisms from A to B or with domain A and range B (we write $\alpha: A \to B$ or $A \xrightarrow{\alpha} B$ for $\alpha \in [A, B]$), these sets being pairwise disjoint: $(A, B) \neq (A', B')$ implies $[A, B] \cap [A', B'] = \phi$;

(iii) for each triple (A, B, C) of objects a map

$$[A, B] \times [B, C] \longrightarrow [A, C]$$
$$(\alpha, \beta) \rightsquigarrow \beta\alpha$$

called composition of morphisms;

(iv) for each object A an element $1_A \in [A, A]$, called identity morphisms;

these data being subject to the two axioms

(1) If $\alpha \in [A, B]$, $\beta \in [B, C]$, $\gamma \in [C, D]$, then $\gamma(\beta\alpha) = (\gamma\beta)\alpha$

(2) If $\alpha \in [A, B]$, then $\alpha 1_A = \alpha$, $1_B \alpha = \alpha$.

Remark. The morphism 1_A whose existence is required by (iv) is uniquely defined by condition 2. Because if $1'_A$ is a second morphism with the same properties, then $1_A' 1_A = 1_A' = 1_A$.

Examples. The category Ens whose objects are the sets and morphisms the maps between sets with the usual compositions. The category $\mathfrak{G}$ of groups is defined by the groups as objects, group homomorphisms as morphisms and the usual composition of homomorphisms. Taking the topological spaces as objects and the continuous maps as morphisms with the usual composition, we obtain the category $\mathfrak{T}$ of topological spaces. Similarly the category $\mathfrak{M}$ of differentiable manifolds is defined by taking the differentiable manifolds as objects and differentiable maps as morphisms.

Let $\mathfrak{K}$ be a category and A, B objects of $\mathfrak{K}$. A morphism $\alpha: A \to B$ is called <u>an equivalence</u> or an isomorphism, if there exists $\beta: B \to A$ with $\beta\alpha = 1_A$ and $\alpha\beta = 1_B$. If there exists an equivalence $\alpha: A \to B$, then A and B are said to be equivalent or isomorphic: $A \cong B$. An automorphism of A is an equivalence $\alpha: A \to A$.

Definition: Let $\mathfrak{K}$ and $\mathfrak{K}'$ be categories. A <u>covariant functor</u> $F: \mathfrak{K} \to \mathfrak{K}'$ from $\mathfrak{K}$ to $\mathfrak{K}'$ is the assignment

(i) of an object FA of $\mathfrak{K}'$ to each object A of $\mathfrak{K}$;

(ii) of a morphism $F\alpha: FA \to FB$ of $\mathfrak{K}'$ to each morphism $\alpha: A \to B$ of $\mathfrak{K}$;

subject to the two conditions

(1) $F(1_A) = 1_{FA}$

(2) $F(\beta\alpha) = F(\beta)F(\alpha)$

If the condition (2) is replaced by

(2') $F(\beta\alpha) = F(\alpha)F(\beta)$,

we speak of a <u>contravariant functor</u> $F: \mathfrak{K} \to \mathfrak{K}'$.

Examples. Let $\mathfrak{K}$ be a category and A an object of $\mathfrak{K}$. One can define a covariant functor $h_A : \mathfrak{K} \to Ens$ in the following way: $h_A(X) = [A, X]$ for any object X of $\mathfrak{K}$, $h_A(\varphi)(\alpha) = \varphi\alpha$ for $\varphi: X \to X'$, $\alpha: A \to X$. Here we have $h_A(\varphi): [A,X] \to [A,X']$. Similarly we can define a contravariant functor $h^A : \mathfrak{K} \to Ens$ by $h^A(X) = [X, A]$ for any object X of $\mathfrak{K}$ and $h_A(\varphi)(\alpha) = \alpha\varphi$ for $\varphi: X \to X'$, $\alpha: X' \to A$. Here we have $h^A(\varphi): [X', A] \to [X, A]$.

Definition. Let $\mathfrak{K}$ and $\mathfrak{K}'$ be categories and $F, G: \mathfrak{K} \to \mathfrak{K}'$ (covariant) functors from $\mathfrak{K}$ to $\mathfrak{K}'$. A natural transformation $\phi: F \to G$ from F to G is the assignment of a morphism $\phi_X: FX \to GX$ to each object X of $\mathfrak{K}$, such that the following diagram commutes for every $\varphi: X \to Y$

$$\begin{array}{ccc} FX & \xrightarrow{\phi_X} & GX \\ {\scriptstyle F(\varphi)}\downarrow & & \downarrow{\scriptstyle G(\varphi)} \\ FY & \xrightarrow{\phi_Y} & GY \end{array}$$

Example. Let K be a commutative field, ${}_K\mathfrak{M}$ the category of K-vectorspaces and $D: {}_K\mathfrak{M} \to {}_K\mathfrak{M}$ the functor defined by the duality (assigning to each vectorspace X its dual space X' and to each K-linear map its dual map). The functor $D^2: {}_K\mathfrak{M} \to {}_K\mathfrak{M}$ assigns to each vectorsubspace over K its bidual. The evaluation $\epsilon_X(x)(x') = \langle x, x' \rangle$ for $x \in X$, $x' \in X'$ defines a natural transformation $\epsilon: 1_{{}_K\mathfrak{M}} \to D^2$.

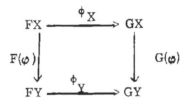

Natural transformations are composed in an obvious way.

A natural transformation $\phi: F \to G$ is a natural equivalence if there exists a natural transformation $\Psi: G \to F$ such that $\Psi\phi = 1_F$, $\phi\Psi = 1_G$, 1_F and 1_G denoting the identical natural transformations $F \to F$ and $G \to G$ respectively.

Product and sums. Let $\mathfrak{K}$ be a category. An object T of $\mathfrak{K}$ is terminal, if to each object K there is exactly one morphism $K \to T$. Hence the only morphism $T \to T$ is 1_T, and any two terminal objects in $\mathfrak{K}$ are equivalent.

Let $(K_j)_{j \in \mathfrak{J}}$ be a family of objects of $\mathfrak{K}$ indexed by a set $\mathfrak{J}$. Consider the category $\mathfrak{P}(K_j)$ whose objects are indexed families $\{q_j: Q \to K_j / j \in \mathfrak{J}\}$ of morphisms of K with a common domain Q, while a morphism $(q_j) \to (q_j')$ in $\mathfrak{P}(K_j)$ is an $\alpha: Q \to Q'$ for which $q_j'\alpha = q_j$ for $j \in \mathfrak{J}$. A terminal object in $\mathfrak{P}(K_j)$ is a product of the K_j, thus

DEFINITION. A product of $(K_j)_{j \in \mathfrak{J}}$ is an object P of $\mathfrak{K}$ together with morphisms $p_j: P \to K_j$ for $j \in \mathfrak{J}$, such that any family of morphisms $q_j: Q \to K_j$ can be written as $q_j = p_j\alpha$ for a unique $\alpha: Q \to P$. The product, like any terminal object, is unique up to an equivalence in $\mathfrak{P}(K_j)$. In particular, the product-object P is unique up to an equivalence in $\mathfrak{K}$.

Let $\mathfrak{K}$ be a category. An object S of $\mathfrak{K}$ is initial, if to each object K there is exactly one morphism $S \to K$. Hence the only morphism $S \to S$ in $1_{\mathfrak{J}}$, and any two initial objects are equivalent.

Let $(K_j)_{j \in \mathfrak{J}}$ be a family of objects of $\mathfrak{R}$ indexed by a set $\mathfrak{J}$. Consider the category $\Gamma(K_j)$ whose objects are indexed families $\{\rho_k : K_j \to R / j \in \mathfrak{J}\}$ of morphisms of $\mathfrak{R}$ with common range R, while a morphism $(\rho_j) \to (\rho_j')$ in $\Gamma(K_j)$ is an $\alpha: R \to R'$ for which $\alpha\rho_j = \rho_j'$ for $j \in \mathfrak{J}$. An initial object in this category is a sum of the K_j, thus

DEFINITION. A <u>sum</u> of $(K_j)_{j \in \mathfrak{J}}$ is an object S of $\mathfrak{R}$ together with morphisms $\sigma_j : K_j \to S$ for $j \in \mathfrak{J}$, such that any family of morphisms $\rho_j : K_j \to R$ can be written as $\alpha\sigma_j = \rho_j$ for a unique $\alpha: S \to R$.

The sum is unique up to an equivalence in $\Gamma(K_j)$, in particular, the sum-object is unique up to an equivalence in $\mathfrak{R}$.

BIBLIOGRAPHY

[1] Bruhat, F., Algèbres de Lie et groupes de Lie, Textos de Mathematica, Univ. do Recife, Vol. 3 (1961).

[2] Bruhat, F., Lectures on Lie groups and representations of locally compact groups. Tata Institute of fundamental research, Bombay, 1958.

[3] Chevalley, C., Theory of Lie Groups, Vol. I, Princeton Univ. Press, Princeton, N. J. (1946).

[4] Cohn, P. M., Lie Groups. Cambridge Univ. Press, Cambridge (1957).

[5] Graeub, W., Liesche Gruppen und affin zusammenhängende Mannigfaltigkeiten, Acta Math. 106 (1961), 65-111.

[6] Helgason, S., Differential geometry and symmetric spaces, Academic Press (1962).

[7] Hoffman, K. H., Einführung in die Theorie der Liegruppen, Teil I. Vorlesungsausarbeitung, Math. Jnst. Universität Tübingen (1963).

[8] Koszul, J. L., Exposés sur les spaces homogènes symétriques. Soc. Math. São Paulo (1959).

[9] Lichnerowicz, A., Géometrie des groupes de transformations, Dunod, Paris (1958).

[10] Maissen, B., Lie-Gruppen mit Banachräumen als Parameterräume, Acta Math. (1962) 229-269.

[11] Nomizu, K. and Kobayashi, S., Foundations of differential geometry, Vol. I., Interscience, N. Y. (1963).

[12] Palais, R., The classification of G-spaces, Memoirs AMS, Vol. 36 (1960).

[13] Palais, R., A global formulation of the Lie theory of transformation groups, Memoirs AMS, Vol. 22 (1957).

[14] Pontrjagin, L. S., Topologische Gruppen, Vol. II, Teubner Leipzig (1958).

B e s c h a f f e n h e i t d e r M a n u s k r i p t e

Die Manuskripte werden photomechanisch vervielfältigt; sie müssen daher in sauberer Schreibmaschinenschrift geschrieben sein. Handschriftliche Formeln bitte nur mit schwarzer Tusche oder roter Tinte eintragen. Korrekturwünsche werden in der gleichen Maschinenschrift auf einem besonderen Blatt erbeten (Zuordnung der Korrekturen im Text und auf dem Blatt sind durch Bleistiftziffern zu kennzeichnen). Der Verlag sorgt dann für das ordnungsgemäße Tektieren der Korrekturen. Falls das Manuskript oder Teile desselben neu geschrieben werden müssen, ist der Verlag bereit, dem Autor bei Erscheinen seines Bandes einen angemessenen Betrag zu zahlen. Die Autoren erhalten 25 Freiexemplare.

Manuskripte, in englischer, deutscher oder französischer Sprache abgefaßt, nimmt Prof. Dr. A. Dold, Mathematisches Institut der Universität Heidelberg, Tiergartenstraße oder Prof. Dr. B. Eckmann, Eidgenössische Technische Hochschule, Zürich, entgegen.

Cette série a pour but de donner des informations rapides, de niveau élevé, sur des développements récents en mathématiques, aussi bien dans la recherche que dans l'enseignement supérieur. On prévoit de publier

1. des versions préliminaires de travaux originaux et de monographies
2. des cours spéciaux portant sur un domaine nouveau ou sur des aspects nouveaux de domaines classiques
3. des rapports de séminaires
4. des conférences faites à des congrès ou des colloquiums

En outre il est prévu de publier dans cette série, si la demande le justifie, des rapports de séminaires et des cours multicopiés ailleurs qui sont épuisés.

Dans l'intérêt d'une grande actualité les contributions pourront souvent être d'un caractère provisoire; le cas échéant, les démonstrations ne seront données qu'en grande ligne, et les résultats et méthodes pourront également paraître ailleurs. Par cette série de »prépublications« les éditeurs Springer espèrent rendre d'appréciables services aux instituts de mathématiques par le fait qu'une réserve suffisante d'exemplaires sera toujours à disposition et que les intéressés pourront plus facilement être atteints. Les annonces dans les revues spécialisées, les inscriptions aux catalogues et les copyrights faciliteront pour les bibliothèques mathématiques la tâche de dresser une documentation complète.

P r é s e n t a t i o n d e s m a n u s c r i t s

Les manuscrits étant reproduits par procédé photomécanique, doivent être soigneusement dactylographiés. Il est demandé d'écrire à l'encre de Chine ou à l'encre rouge les formules non dactylographiées. Des corrections peuvent également être dactylographiées sur une feuille séparée (prière d'indiquer au crayon leur ordre de classement dans le texte et sur la feuille), la maison d'édition se chargeant ensuite de les insérer à leur place dans le texte. S'il s'avère nécessaire d'écrire de nouveau le manuscrit, soit complètement, soit en partie, la maison d'édition se déclare prête à se charger des frais à la parution du volume. Les auteurs reçoivent 25 exemplaires gratuits.

Les manuscrits en anglais, allemand ou français peuvent être adressés au Prof. Dr. A. Dold, Mathematisches Institut der Universität Heidelberg, Tiergartenstraße ou Prof. Dr. B. Eckmann, Eidgenössische Technische Hochschule, Zürich.

Printed in the United States
By Bookmasters